AF691180

Electricity in Economic Development

Recent Titles in
Contributions in Economics and Economic History

The Coming of Age of Political Economy, 1815–1825
Gary F. Langer

The Reconstruction of Economics: An Analysis of the Fundamentals of Institutional Economics
Allan G. Gruchy

Development Finance and the Development Process
Kempe Ronald Hope

Bridging the Gap Between Rich and Poor: American Economic Development Policy Toward the Arab East, 1942–1949
Nathan Godfried

Rethinking the Nineteenth Century: Contradictions and Movements
Francisco O. Ramirez, editor

Textiles in Transition: Technology, Wages, and Industry Relocation in the U.S. Textile Industry, 1880–1930
Nancy Frances Kane

Threats of Quotas in International Trade: Their Effect on the Exporting Country
Gerard Lawrence Stockhausen

A Slippery Slope: The Long Road to the Breakup of AT&T
Fred W. Henck and Bernard Strassburg

The Suppression of the Automobile: Skulduggery at the Crossroads
David Beasley

New Perspectives on Social Class and Socioeconomic Development in the Periphery
Nelson W. Keith and Novella Zett Keith, editors

World Population Trends and Their Impact on Economic Development
Dominick Salvatore, editor

Racism, Sexism, and the World-System
Joan Smith, Jane Collins, Terence K. Hopkins, and Akbar Muḥammad, editors

ELECTRICITY IN ECONOMIC DEVELOPMENT

The Experience of Northeast Asia

Edited by
Yoon Hyung Kim and Kirk R. Smith
with the assistance of
Kennon Breazeale

Resource Systems Institute
East-West Center, Honolulu, Hawaii

Contributions in Economics and Economic History, Number 85
Robert Sobel, Series Editor

GREENWOOD PRESS
New York • Westport, Connecticut • London

Library of Congress Cataloging-in-Publication Data

Electricity in economic development : the experience of Northeast Asia / edited by Yoon Hyung Kim and Kirk R. Smith, with the assistance of Kennon Breazeale.
p. cm. — (Contributions in economics and economic history, ISSN 0084-9235 ; no. 85)
Bibliography: p.
Includes index.
ISBN 0-313-26228-4 (lib. bdg. : alk. paper)
1. Electric power consumption—East Asia. 2. East Asia—Economic conditions. I. Kim, Yoon Hyung. II. Smith, Kirk R. III. Breazeale, Kennon. IV. Series.
HD9685.A764E44 1989
333.79'3213'095—dc 19 88-9754

British Library Cataloguing in Publication Data is available.

Copyright © 1989 by Resource Systems Institute, East-West Center

All rights reserved. No portion of this book may be reproduced, by any process or technique, without the express written consent of the publisher.

Library of Congress Catalog Card Number: 88-9754
ISBN: 0-313-26228-4
ISSN: 0084-9235

First published in 1989

Greenwood Press, Inc.
88 Post Road West, Westport, Connecticut 06881

Printed in the United States of America

The paper used in this book complies with the Permanent Paper Standard issued by the National Information Standards Organization (Z39.48-1984).

10 9 8 7 6 5 4 3 2 1

Dedicated to

Harrison Brown (1917–1986)
Mentor, Colleague, and Friend

Contents

Tables

FIGURES

Preface

This historical study is a product of a cost-shared and region-to-region collaborative research effort between the East-West Center and a group of institutions in China (Taipei), Japan, and Korea, coordinated by the East-West Center, in consultation with the Massachusetts Institute of Technology (MIT) and the Electric Power Research Institute (EPRI) in California.

Recognizing the critical role of the electric power sector in influencing a nation's future industrial development and its energy supply-demand pattern in the post-oil crisis era, Dr. Harrison Brown, then director of the East-West Center's Resource Systems Institute, initiated in 1981 a project entitled The Electric Future of the Asia-Pacific Region within the Energy Program of the East-West Center. Dr. Brown also launched this project in the light of the coming Pacific Rim Era based on his far-sighted vision that the winds of cultural gravity in world history, having blown from the Mediterranean to the Atlantic, will blow from the Atlantic to the Pacific in the twenty-first century. In view of population distribution, resource endowments, and industrialization potentials, trade activities will become even more dynamic in Pacific Rim countries in the next century. It is hoped that information provided in this assessment will stimulate and advance regional cooperation on energy trade, development, demand management, financing, and efficient modes of transportation, arising from mass consumption of coal, liquified natural gas, and uranium in the post-oil crisis era.

In assessing the electric future of Northeast Asia, Dr. Brown successfully pulled together the accumulated practical expertise of China (Taipei), Japan, and Korea and the technical expertise of the United States (MIT and EPRI) at the East-West Center as a regional intellectual vehicle for bridging region-to-region cooperation in the Pacific Rim countries. The project has been conducted on a cooperative basis with Taiwan Power Company, the Council for Economic Planning and Development, and the Energy Committee, Ministry of Economic Affairs in China (Taipei); with the Japanese Task Force, including members from the Tokyo Electric Power Company, the Central Research Institute of Electric Power Industry, the Institute of Energy Economics, and the Mitsubishi Research Institute; and with the Korea Electric Power Corporation.

This project is the first attempt to mobilize senior power experts in all three countries of Northeast Asia for cooperative electric future assessment and to disseminate within the Asia-Pacific region the information and lessons provided. The electric power industry in any country is so dynamic in its activities and so extensive in its coverage that integrated electric power data compiled in a consistent way for policy analysis have not previously been made ready for outside researchers. Accordingly, it is hoped that this work will serve as a valuable reference for energy policy making, enabling the countries of the Asia-Pacific region to improve their energy decision making in the future by making cross-country comparisons and acquiring clear insights into what has happened in the three Northeast Asian power industries in the past.

The East-West Center acknowledges the full support provided by the numerous institutions that have collaborated in this project.

ACKNOWLEDGMENTS

The following is a list of the principal contributors and their areas of expertise:

CHINA (TAIPEI)

Mr. Wen-Po Ke Taiwan Power Company	Load management, tariffs, financing
Mr. Kao-Chao Lee Council for Economic Planning and Development	Demand analysis
Mr. Shuo-Chung Li Ministry of Economic Affairs	Demand analysis
Mr. Eng Lin Taiwan Power Company	Nuclear issues
Mr. Pu-Sen Pan Taiwan Power Company	Supply analysis
Mr. Hui-Sheng Wang Taiwan Power Company	Supply analysis
Mr. Tyng-Sheng Yang Taiwan Power Company	Demand analysis

JAPAN

Mr. Toru Kimura Institute of Energy Economics	Demand and supply analysis
Mr. Kazutoshi Kuwahara Industrial Bank of Japan, Ltd.	Nuclear analysis
Mr. Tomoyoshi Sawada Tokyo Electric Power Company	Supply analysis
Mr. Akira Tani Mitsubishi Research Institute	Demand analysis
Mr. Mitsuho Uchida Central Research Institute of Electric Power Industry	Financing and tariffs

KOREA

Mr. Eung-Wan Kim Korea Electric Power Corporation	Financing and tariffs
Dr. Ho-Tak Kim Seoul National University	Demand and supply analysis
Mr. Han-Kwae Lim Korea Electric Power Corporation	Nuclear issues
Mr. Chung Il Nahm Korea Electric Power Corporation	Demand and supply analysis
Mr. Mahn-Cherl Shin Korea Electric Power Corporation	Supply analysis

UNITED STATES

Dr. Manuel Castillo-Bonet Massachusetts Institute of Technology	Supply analysis
Dr. Hung-Po Chao Electric Power Research Institute	Modeling
Mr. Tsyr-Hsioung Lee Electric Power Research Institute	Demand analysis
Dr. David Rose Massachusetts Institute of Technology	Senior adviser and nuclear issues
Dr. Richard Tabors Massachusetts Institute of Technology	Load management, tariffs, financing
Dr. Oliver Yu Electric Power Research Institute	Demand analysis

Special appreciation is due Dr. Ken Breazeale of the Resource Systems Institute for his assistance during all phases of the project and book preparation. We also gratefully thank the graduate students and research interns who helped prepare tables and figures: Stuart Cooley, Yoshiki Iinuma, George J. Y. Hsu, and Allen Jelich. We appreciate the professional support provided by East-West Center staff: Jackie D'Orazio, David Puhlick, April Kam, and Audrey Shono. We appreciate the continued support of the Resource Systems Institute from Dr. Seiji Naya, Director, and Dr. Fereidun Fesharaki, Energy Program leader.

Most of the data used in the preparation of this book came directly from the utility records of the three countries. As a result, they often cannot be cited in the same way as data from published literature. Instead, we have chosen to list in the selected bibliography the working papers and other documents in which the original data were first compiled as part of the collaborative research project.

We apologize to readers who may find the terminology used in this book for China to be awkward and distracting. We adopted the terms China (Taipei) and China (Beijing) at the start of the project because they were among the few acceptable to both at that time.

Y. H. Kim
K. R. Smith
Resource Systems Institute
East-West Center, Honolulu, Hawaii

ABBREVIATIONS

A/E	Architecture/Engineering
AEC	Atomic Energy Commission
AFR	away from reactor
ASEAN	Association of Southeast Asian Nations
ATR	advanced thermal reactor
BWR	boiling water reactor
c-km	circuit-kilometers
CAS	Commission on Assurance and Supply
DSS	demand subscription service
DOE	Department of Energy
EEI	Edison Electric Institute
EPDC	Electric Power Development Co.
EPRI	Electric Power Research Institute (California)
ESCAP	Economic and Social Commission for Asia and the Pacific
FEPC	Federation of Electric Power Companies
GDP	gross domestic product
GNP	gross national product
GWh	gigawatt-hour = one million kilowatt-hours
GW	gigawatt = one million kilowatts
HTGR	high temperature gas-cooled reactor
HWR	heavy water reactor
IAEA	International Atomic Energy Agency
INFCE	International Fuel Cycle Evaluation
INPO	Institute of Nuclear Power Operations
IPS	international plutonium storage
ISWACO	Industrial Site and Water Development Co.
JAERI	Japan Atomic Energy Research Institute
JAPCO	Japan Atomic Power Co.
JEPSC	Japan Electric Power Survey Committee
JNFS	Japan Nuclear Fuel Service Co.
KAEC	Korea Atomic Energy Commission
KAERI	Korea Advanced Energy Research Institute
KECO	Korea Electric Power Co.

KEPCO	Korea Electric Power Corporation
kl	kiloliter
km	kilometer
KNU	Korean nuclear unit
KOPEC	Korea Power Engineering Co.
kW	kilowatt
KWU	Kraftwerke Union
LDC	less developed country
LNG	liquified natural gas
LOLP	loss of load probability
LPG	liquified petroleum gas
LRMC	long-run marginal costs
LWR	light water reactor
MIT	Massachusetts Institute of Technology
MITI	Ministry of International Trade and Industry
MVA	Megavolt-Amps
MW	Megawatt
MWth	Megawatts thermal
NIC	newly industrializing country
NPT	Non-Proliferation Treaty
NRB	Nuclear Regulatory Bureau
NRD	Nuclear Regulatory Division
NSAC	Nuclear Safety Analysis Center
NSC	Nuclear Safety Commission
OECD	Organisation for Economic Co-operation and Development
OPEC	Organization of Petroleum Exporting Countries
PECI	Pacific Engineers and Constructors Inc.
PECO	Philadelphia Electric Co.
PG&E	Pacific Gas and Electric Company
PRIS	power reactor information system
PWR	pressurized water reactor
R&D	research and development
SRMC	short-run marginal cost
T&D	transmission and distribution
TEPCO	Tokyo Electric Power Co.
TMI	Three Mile Island
USNRC	U. S. Nuclear Regulatory Commission

Electricity in Economic Development

CHAPTER 1

ELECTRICITY IN NORTHEAST ASIA

The Principal Historical Trends

Y. H. Kim and K. R. Smith

Widespread and intensive electricity consumption is a foundation of modern society. As time has gone on, increasing fractions of primary energy consumption in all societies have gone into its generation. The use of electricity on a really large scale, however, is of recent origin, although its use began more than a century ago with the first central power station built by Thomas Edison in 1882.

There have been many growing pains in the industry in all countries, including those of standardization, the shift from direct to alternating current, and the recognition of critical problems such as transmission losses over long distances. Serious management problems appear to have been due to the need to smooth the uneven load and finance the hugely expensive systems. In the United States after World War I, for example, the inventor-entrepreneurs were replaced by financial managers and engineers as the prime controllers of the system—a necessary step in the creation of the large interconnected regional networks.

By 1980 annual electricity consumption in the United States had exceeded 10,000 kilowatt hours per inhabitant and was still rising. Consumption in the other industrialized countries was climbing rapidly, probably eventually to reach or exceed the U.S. level. Although electrical consumption is closely related to the economic activity of a country, it is difficult to forecast the ultimate level of consumption. Largely because of the convenience of electricity, it is quite possible, however, that someday we will reach the point where the greater part of our energy consumption passes through the electrical route and we will have what amounts to an electrified society.

For the past several decades there has been a strong relationship between the electrical consumption of a country and its level of economic activity, as measured by an indicator such as the gross national product. In 1980 the ratio of GNP to the electrical power generated in the United States was close to US$1 per kilowatt hour. The average for Asia as a whole was about US$1.20 per kilowatt hour, including India (US$0.96) and Japan (US$1.22). Although this ratio is an interesting and useful indicator, it is by no means an invariant. It depends markedly upon the nature of the economic activities, upon the efficiency of the electrical network, and upon the effectiveness of the utility management in controlling costs and loads.

In the early 1970s it appeared that humanity was truly fissioning into two groups of nations: the rich and the poor. We see this when we examine the distribution of per capita energy consumption, where there is a marked clustering around a high level of consumption and another around a low level of consumption. We find the same separation when we examine the consumption of electricity, steel, and the distribution of the gross national products.

When we examine the changes of these variables with time, we see that they appear to increase over the years in the rich nations and the poor nations alike. The data for the period 1970–1980, however, show at least the beginnings of a major change. Per capita consumption in the rich nations seems to be now growing relatively slowly, whereas per capita consumption in the poor nations is still growing rapidly, although somewhat more slowly than it did in the 1970s. From 1970 to 1980 commercial energy consumption in the rich nations grew at a rate of about 0.4 percent per year; that in the poor nations grew at a rate of 5.1 percent per year. The disparities in the growth of electricity consumption were equally prominent.

A more detailed look at these trends shows that the developing countries are divided basically into three groups. The bulk of the population, including India and China (Beijing), is experiencing growth at similar or somewhat higher rates than developed countries. Another group, mainly in Africa, is experiencing little or even negative growth.

A significant part of this consumption increase in the developing world has occurred in the third group of nations: newly industrializing countries, or NICs. Their economies have grown much faster than the other developing countries and, indeed, have been rapidly crossing the gap between the poor and rich nations of the world. Some of the relevant characteristics of the major NICs are compared with representative rich and poor nations in Table 1.1. Obviously, many other developing countries would like to follow a similar path.

Northeast Asia boasts the two most successful NICs of the immediate pre- and post-energy crises period (1960–1982). During these years, the economies of China (Taipei) and the Republic of Korea have grown faster than that of any other nonoil-exporting country of any size. Their neighbor, Japan, might once have been classified as a NIC, even though it achieved its major industrialization much earlier. It has served as a model to its neighbors, however, due to its spectacular ability to create and manage rapid economic growth.

In common with many developing countries, the rate of growth of electri-

Table 1.1
Economic Indicators in Selected Countries

Country	Population		GNP/Capita		Industry		Energy		Electricity	
	1982 (millions)	1960–1982 (%)	1982 (US$ 1982)	1960–1982 (%)	1960 (% of GDP)	1982 (% of GDP)	1982 (kg/capita coal equivalent)	1960–1982 (%)	1982 (kWh/ capita)	1960–1982 (%)
Industrial market										
Australia	15.2	1.7	11,140	2.4	40	35	7,164	3.5	6,934	5.3
Japan	118.4	1.1	10,080	6.1	45	42	3,503	5.7	4,916	6.5
United States	231.5	1.1	13,160	2.2	38	33	9,431	0.7	10,074	3.5
Newly industrializing										
Brazil	126.8	2.6	2,240	4.8	35	35	704	3.9	1,196	6.3
China (Taipei)	18.5	2.1	2,380	6.7	30	49	2,218	6.5	2,184	9.3
Hong Kong	5.2	2.4	5,340	7.0	39	30	1,694	6.2	2,697	8.8
Korea	39.3	2.1	1,910	6.6	20	39	1,442	9.3	1,204	13.8
Mexico	73.1	3.1	2,270	3.7	29	38	1,760	4.2	1,101	6.0
Yugoslavia	22.6	0.9	2,800	4.9	45	45	2,331	4.7	2,833	8.4
Middle-income										
Indonesia	152.6	2.2	580	4.2	14	39	229	3.0	47	5.3
Malaysia	14.5	2.6	1,860	4.3	18	30	984	4.8[a]	850	8.3[a]
Philippines	50.7	2.8	820	2.8	28	36	329	4.1	382	6.4
Thailand	48.5	2.7	790	4.5	19	28	364	8.5	370	13.7
Low-income										
Bangladesh	92.9	2.6	140	0.3	7	14	50	—	35	—
India	717.0	2.3	260	1.3	20	26	200	2.9	193	6.6
Nepal	15.4	2.3	170	-0.1	—	—	11	4.7	17	13.7
Sri Lanka	15.2	2.0	320	2.6	20	27	120	0.7	134	6.9

Sources: World Development Report (1984); UN Energy Statistics Yearbook (1982).

a. 1970–1982.

— Not available.

cal consumption in the Northeast Asian NICs has been greater than that of total energy consumption, which is greater than the growth rate of the urban population, which is greater than the rate of growth of the population nationwide. If the history of economic development provides a valid guide, this trend will continue until such time as the countries are virtually completely urbanized, population ceases to grow, energy needs per capita level off, and electricity has penetrated the energy market to the extent found in developed countries.

Each of these countries has faced tremendous difficulties in managing their electric power sectors during rapid economic growth. Electric power consumption grew at annual rates above that of the economy or of energy consumption as a whole and well into double digits for decades during this period. Generating capacity increased by more than a factor of ten. All this occurred during times of drastically changing global pictures for energy supplies and capital flows as well as major innovations in technology and management for the power industry.

It seems likely that Japan and the United States will shortly equal each other in per capita electrical generating capacity. It also seems likely that China (Taipei) and Korea will eventually reach equally high levels. At present there appears to be about a 15-year lag between the development of China (Taipei) and Japan. The lag between Korea and Japan appears to be about 20 years.

One of the principal development uncertainties today is whether the NICs can continue their economic performance and whether other developing countries can follow. The Southeast Asian countries such as Thailand and the Philippines, for example, have many economic characteristics that are similar to those of the NICs before their great growth spurts. As in the present NICs, however, good management of the power sector will be critical to achieving this goal, because of its huge demand for capital and primary energy as well as the critical importance of its unique product for the economy.

This book examines the approaches of the electric utilities in Northeast Asia in meeting the challenges posed by the rapid changes in electrical demand and the world energy scene. Its objective is to achieve an improved understanding of the factors that determine electricity demand and the influence of alternate modes of generation, distribution, and management in meeting that demand. A historical and comparative approach has been taken, taking advantage of the common cultural heritage and geographical situation of the three countries.

Although these countries started the development process at widely differing times, their experiences have nevertheless been remarkably similar. We believe that a detailed comparison of those experiences can be useful to development planners, no matter where they live, but particularly to those who are planning the electrical networks of other countries anticipating rapid economic growth.

THE MATURATION OF ENERGY POLICY

Until the 1970s, the main concern of energy policy was to ensure that the availability of energy did not become an obstacle to economic growth. Since the overall development strategy in most countries was to build a strong industrial base, most efforts were directed toward providing a stable supply of cheap energy to the critical sectors of the economy. Thus, energy played a supportive role in a country's industrialization. In this situation, energy policy was limited simply to adjusting supply to demand and did not require sophisticated analysis.

This type of energy policy was made possible by the continuous availability of relatively inexpensive imported oil supplies, so that energy constraints on economic growth were minimal until the increase of world oil prices in 1973. The lack of proper characterization of energy as a factor in production is not surprising considering that, prior to the autumn of 1973, the scarcity value of energy products received scant attention. Energy was a topic of physicists, not of planners and decision makers. The manyfold increase in the real price of crude oil since 1973, however, has elevated energy expenses to a significant portion of production costs in many industries, and hence it has raised disturbing questions about the prospects of industrialization. The direct and indirect energy cost of producing many outputs has risen to the point where this cost presumably will become a principal determinant of the location of world productive capacity. Although reducing the urgency, the lower prices of the mid-1980s have not led most countries to abandon concerns over long-term energy prices and supply.

The price and availability of energy having become significant constraints on world production, consumption, and trade since 1973 necessitates a new approach to economic development planning—one that takes more explicit account of the role of energy. Because energy pervades all aspects of economic activity and its demand is derived from the structure and growth of the whole economy, the policy goal of achieving an optimal balance of supplies and conservation of energy can be realized only if supported by adequate management of demand in the consuming sectors (industry, transportation, residential, and commercial). Therefore, it is essential that the energy implications of alternative development policies in all sectors be properly appraised, and also that effective trade-offs be made between energy efficiency and additional capital expenditures, fuel conversion and additional capital expenditures, and technologies that differ in energy intensity and capital cost. Government regulations, research funding, taxes, and subsidies, which in the past may have favored centralized power production over alternative means of achieving the same result, will have to be balanced against the needs of the entire economy.

Electricity is a high-quality energy form; its versatility, flexibility, adaptability, and amenability to control are unmatched. There are certain applications that require electricity (for example, communication, lighting, and computation). In addition, the cleanliness, convenience, familiarity, and reliability of electricity make it the energy source of choice in many other applications. In industrial usage, electricity can often replace many flame-based technologies, with consequent improvements in product quality. In a growth environment, electricity will further penetrate the chemical, food, metal, and metallurgy industries, including such applications as microwaves, plasma, electric arcs, lasers, and membrane processes.

At present, about a third of total commercial primary energy sources go directly to power generation in most countries. This is expected to increase to as high as 40 to 50 percent by the year 2000. The World Bank estimates the present percent of commercial energy devoted to electricity production to be 25 percent for all developing countries and projects a rise to 30 percent by 1991.

In addition, whereas there are rough correlations between overall energy consumption and economic growth, experience since the oil crises has shown that the linkage is not as strong as once was believed. For the electric power sector, however, the evidence of a strong linkage is more convincing. In a study of 15 OECD countries, for example, the U.S. Electric Power Research Institute (Searl, 1986) found essentially linear relationships between income and electricity consumption in every case, although with widely differing slopes.[1] Even in those countries such as the United States, for example, where a substantial degree of economic growth has occurred without major increases in energy consumption, there still remains a significant growth in power demand. There are even significant exceptions to this linearity, however, particularly at lower levels of economic development. The states of India, many of which are as populous as good-size countries, do not exhibit a close correlation of electricity and income (Pachauri, 1982). In addition, at the other end of the development spectrum there is evidence of a leveling or falling of electricity intensity in the manufacturing sector (i.e., in the United States; Kahane and Squitieri, 1987).

Electric power is perhaps the most important element in the energy sector. This is true not only because power demand typically grows faster than total energy demand, but also because the power sector is usually the largest consolidated end-user of primary energy sources. Moreover, the investment lag in the power sector is the longest and its incremental capital output ratio is the largest in the economy.

This increasing role for electric power in the economy can be partly explained by the rapid flow of technological innovation that has pushed many industries, businesses, and households toward new machines, equipment, and appliances that depend on electricity's distinctive versatility. Whatever the reasons, recent history has shown that it is usually the most rapidly growing energy sector during economic development. Accordingly, the power sector's energy conservation and fuel policies can drastically change a country's overall energy picture. This is well illustrated through a historical analysis of the economy-energy-electricity relationships in Northeast Asia, the region of the world's most rapidly growing economies during the 1960s and 1970s.

THE IMPORTANCE OF ELECTRIC POWER TO DEVELOPMENT

When the time series data of China (Taipei), Japan, and Korea are pooled on a chart showing per capita energy consumption plotted against per capita income, there is a close relationship between consumption of energy and gross domestic product (Figure 1.1).[2] Figure 1.2 shows the 1955–1981 cross-country statistical relationships between per capita electricity consumption and per capita GDP. The relationships again appear to be approximately log-linear.

It is obvious from these figures that China (Taipei) and Korea closely followed Japanese patterns of energy and electricity consumption, although with a time-lag of roughly 15 and 20 years, respectively. The resource endowments, sociocultural heritage, and even the economic and administrative arrangements of China (Taipei) and Korea have much in common with those of Japan. With each step in Japan's industrial evolution from labor-intensive industries to capital-energy-electricity-intensive manufactures, and finally to high-technology industries, its two neighbors have been one or two steps behind. This may help partly explain why China (Taipei) and Korea moved up along the same trend curve established by Japan during the past 25 years. Looking to the future, a key question is whether they will continue to follow the same relationship between per capita GDP and per capita electricity consumption as Japan already experienced. It would be clearly useful for China (Taipei) and Korea to study the Japanese experiences when making long-term electricity plans. It may also be valuable for other developing countries to explore the history of these two highly successful NICs for the purpose of finding pertinent lessons for their own development.

Table 1.2 dramatizes the association even more. From 1960 to 1980 the average annual growth rate of electric power demand (8.7 percent) was substantially higher than either that of total energy demand (7.7 percent) or of GDP (7.6 percent) in Japan. In Korea power demand grew by an average annual rate of 19.1 percent, compared with only 7.7 percent of total energy demand and 8.6 percent of GDP. As did Japan and Korea, China (Taipei) experienced an average annual growth rate of electric power demand above the growth rates of total energy demand and GDP.

The same history for the developing countries of Economic and Social Commission for Asia and the Pacific (ESCAP) is shown in Table 1.3. Between 1973 and 1978, electricity consumption sharply increased at the overall average annual rate of 10.2 percent, whereas GDP and total commercial energy consumption grew by average annual rates of 7.1 and 8.5 percent, respectively.

A large fraction of total energy goes into electricity production in Northeast Asia as shown in Table 1.4. In 1980 the proportion ranged from 22 percent in Korea to 33 percent in the United States, to 35 percent in China (Taipei), and to 36 percent in Japan.

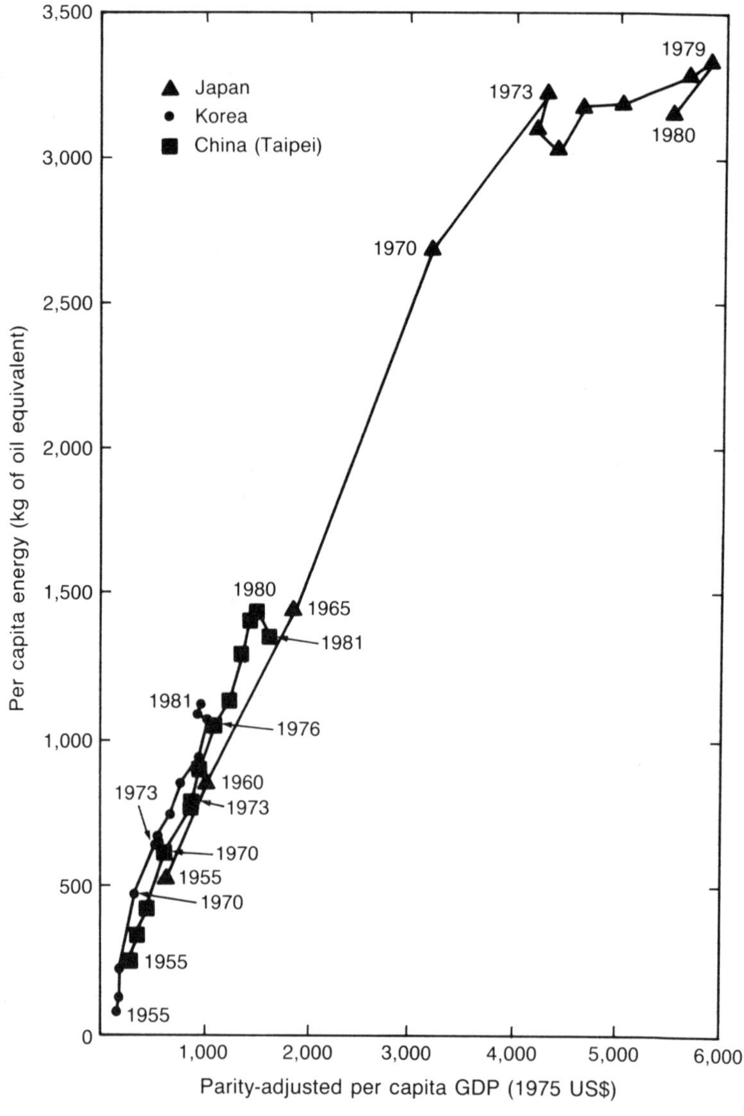

Figure 1.1. Historical energy consumption and GDP in Northeast Asia. Note that China (Taipei) and Korea are following a slightly more energy-intensive route than Japan.

The electric power sector is capital intensive and in 1980 directly accounted for 7 percent of the total fixed capital formation in Japan, 13 percent in Korea, and 17 percent in China (Taipei), as shown in Table 1.5. In particular, in Korea electric power investments amounted to more than two-thirds of the total

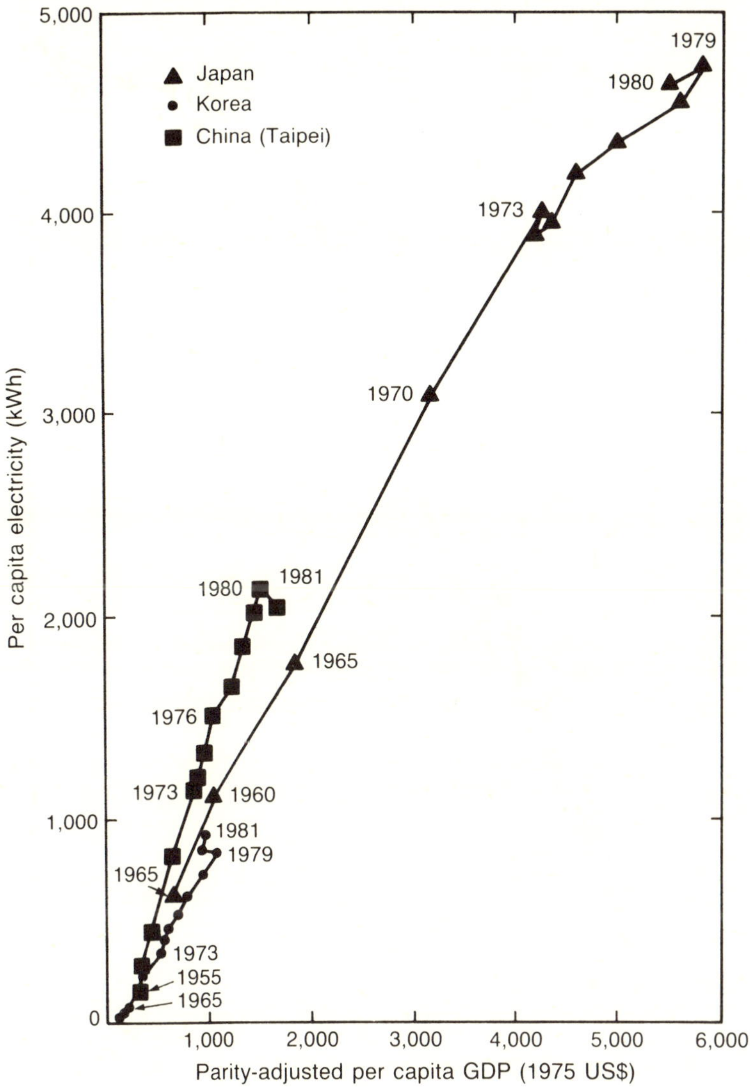

Figure 1.2. Historical electricity consumption and GDP in Northeast Asia. Note that China (Taipei) is following a slightly more electricity-intensive path than either Japan or Korea.

manufacturing investment. The developing countries can be expected to give high priority to the development of electric power as essential infrastructure development and can accordingly be expected to allot substantial financial resources to this sector. It is conceivable that a less power-intensive economy, requiring a smaller percentage of total investment in the power sector, might enable the country to achieve a higher economic growth rate. Consequently, the management and financial performance of the power sector would have a significant influence on the overall capital efficiency of the economy and therefore would help to determine the economic future and well-being of the nation.

ELECTRICITY'S ROLE IN THE ENERGY TRANSITION

With limited domestic energy resources, China (Taipei), Japan, and Korea have become more and more dependent on imported energy sources to meet the rapid increase in energy demand. As Table 1.6 shows, from 1965 to 1980 the proportion of imported energy sources in the total of primary energy consumption rose in all three. More critically, the countries of Northeast Asia depend on imported oil as their major primary source to the extent that it constitutes 60 to 70 percent of the total primary energy consumption. Consequently, they have had to reduce their dependence on imported oil to limit their vulnerabilities to supply interruptions.

Although the oil age is far from over, we must face the harsh reality that future access to oil supplies will be limited by resource constraints in some countries and by economic and political factors in others. Overdependence on oil has already resulted in serious economic dislocations. If this dependence is not reduced, economic difficulties could continue to confront the

Table 1.2
Comparative Evolution of GDP, Energy, and Power Consumption, 1960–1980

Item	China (Taipei)	Japan	Korea[a]
Annual growth rate (%)			
Real GDP	9.6	7.6	8.6
Total energy consumption	10.3	7.7	7.7
Electric power sold	13.8	8.7	19.1
Elasticities			
Energy GDP	1.07	1.01	0.96
Power GDP	1.44	1.14	2.24

Source: Kim (1983).

a. Korean data are for the period 1961–1980.

OECD countries, and slower economic growth with high inflation will affect the developing countries. Yet with current technical capabilities, arising in part from the economic development brought about in the past by cheap oil, a transition to alternative sources of energy can now be achieved. There remain formidable obstacles to a successful energy transition. Some of these are rooted in social and individual attitudes. Others are the results of politi-

Table 1.3
Growth Rates of GDP and Energy Consumption, 1973–1978, by Percentage

Country	GDP	Total Commercial Energy	Electricity Generation
Energy-exporting countries			
Indonesia	6.6	13.5	8.2
Malaysia	7.0	4.7	14.4
	(6.7)[a]	(11.0)	(11.7)
Energy-importing countries			
Low-income group			
Afghanistan	8.3	2.2	5.7
Bangladesh	6.1	12.2	6.2
Burma	5.5	5.9	8.0
Nepal	3.8	-0.4	13.4
Pakistan	4.5	3.5	4.3
Sri Lanka	4.0	1.5	7.3
	(5.2)	(4.4)	(5.1)
Lower middle-income group			
Papua New Guinea	2.6	5.0	9.1
Philippines	6.1	4.3	5.2
Thailand	7.4	6.1	12.5
	(6.6)	(5.2)	(7.9)
High- and upper middle-income group			
China (Taipei)	8.4	11.9	12.2
Fiji	2.6	5.3	5.5
Hong Kong	8.3	12.1	6.0
Korea	9.8	9.7	15.4
Singapore	6.9	5.6	9.6
	(8.7)	(10.1)	(12.3)
	[7.1][a]	[8.5]	[10.2]

Source: Asian Development Bank, *Regional Energy Survey,* March 1981.

a. Numbers in parentheses indicate weighted average; numbers in brackets show weighted average of all countries.

cal, economic, and cultural constraints and the volatile nature of the international oil market that obscures long-term trends.

The three countries have made considerable efforts to decrease their oil dependency, partly through efficient use of energy and partly through fuel substitution. Although the main benefits are expected to accrue during the late 1980s, some of the decisions taken since 1974 have already brought about significant changes in energy supply patterns (also shown in Table 1.6). The share of imported oil in the total primary energy supplies in Japan has decreased since 1973. On the other hand, the share of imported oil in China (Taipei) was reduced only after 1977. Likewise, Korea's oil dependency went on rising until 1979, after which it started to decline.

These energy transitions away from oil in China (Taipei), Japan, and Korea have been achieved mainly through restructuring fuel patterns for power generation. The power sector provides a major vehicle for introducing new energy technologies into practice, manipulating a nation's overall energy strategy, and restructuring its energy supply and import patterns.

Figure 1.3 illustrates this point with respect to the three Northeast Asian countries and the United States. Between 1973 and 1980 the proportion of oil in sectoral primary energy demand by the electric generation sector of Japan declined dramatically, compared with a minor decrease in nongeneration sectors. This decrease in oil dependency, stemming mainly from interfuel substitution in the electric power sector, pulled overall oil dependency downward. The case of the United States dramatizes the contribution of power generation in energy transition. From 1973 to 1981 the electric generation sector reduced its oil dependency, whereas the oil dependency of nongeneration sectors increased slightly. During the same period in Korea the proportion of oil as the primary energy source of the electric generation sector declined somewhat, whereas the proportion in nongeneration sectors continued to increase until 1979. For China (Taipei), even after the price of crude petroleum rose in 1973–1974, the share of oil in total primary energy consumption rose until 1977. Between 1977 and 1980 the electric generation sector's oil dependency sharply declined while that of the nongeneration sectors continued to increase slowly.

Table 1.4
Source Energy for Electricity Generation as a Percentage of Total Commercial Energy Consumption, 1965–1980

Year	China (Taipei)	Japan	Korea	United States
1965	30.7	31.8	14.7	21.0
1975	34.9	32.8	20.8	28.9
1980	34.7	36.1	21.8	33.4

Source: Kim (1983).

Table 1.5
Investment in the Power Sector as a Percentage of Total Fixed Capital Formation, 1976–1980

Country and Sector	1976	1977	1978	1979	1980
China (Taipei)					
Electricity sector[a]	11.3	10.9	14.7	14.9	16.9
Manufacturing	37.0	29.5	24.5	27.5	29.4
Japan					
Electricity sector[a]	6.5	6.6	7.8	6.5	6.9
Manufacturing	22.6	20.6	18.1	20.9	23.4
Korea					
Electricity sector[a]	9.8	9.1	8.4	10.0	12.9
Manufacturing	21.2	19.9	22.6	21.1	17.5

Source: Kim (1983).

a. Includes gas and water.

Emancipation from oil dependence in China (Taipei), Japan, Korea, and, for comparison, the United States is occurring not only at varying rates but also in different forms. As shown in Table 1.6, there have been several significant shifts in source energy for power generation. Between 1973 and 1980 Japan carried out interfuel substitution in the electric generation sector to reduce its oil dependency by sharply increasing the share of nuclear and natural gas. On the other hand, the power sector of the United States managed interfuel substitution mainly by increasing the share of coal and nuclear. As in Japan, the share of nuclear in generation fuel requirements of China (Taipei) and Korea has risen substantially, starting at virtually nothing in 1977. Unlike Japan, coal consumption in China (Taipei) and Korea has also shown a substantial rise.

Interfuel substitution away from oil in the electric power sector has been one of the keys to a successful energy transition in China (Taipei), Japan, Korea, and even to some extent, the United States. This is so even though none of the Northeast Asian countries has significant domestic energy resources. In the future, a continually greater proportion of total commercial primary energy consumption will be used to produce electricity (see Table 1.4). The development of electricity by means other than burning fuel oils will offer one of the more promising long-term prospects for oil-deficient countries to improve their energy situations. Accordingly, the electric power sector will continue to be an engine of energy transition in the postpetroleum era.

Table 1.6
Structure of Primary Energy Demand by End-Use Sector, 1965–1980, by Percentage

Energy-Consuming Sector and Year	Petroleum	Coal	Natural Gas	Hydro	Nuclear	Firewood
China (Taipei)						
Power generation						
1965	12	46	3	39	0	—
1969	54	11	9	27	0	—
1975	67	8	3	27	0	—
1977	79	6	2	13	0.3	—
1980	58	16	0	7	19	—
Nongeneration[a]						
1965	34	60	6	0	0	—
1969	44	47	10	0	0	—
1975	69	18	13	0	0	—
1977	74	14	13	0	0	—
1980	77	13	9	0	0	—
Total						
1965	27	56	5	12	0	—
1969	47	35	10	8	0	—
1975	69	15	9	8	0	—
1977	76	11	9	4	0.1	—
1980	70	15	6	2	7	—
Japan						
Power generation						
1965	31	28	0.4	39	0	0
1973	71	9	2	16	2	0
1975	63	8	4	19	6	0
1978	58	7	10	14	11	0
1980	46	9	13	17	15	0
Nongeneration[a]						
1965	66	3	2	0	0	3
1973	79	19	2	0	0	0.2
1975	77	21	2	0	0	0.4
1978	81	16	3	0	0	0
1980	76	21	3	0	0	0
Total						
1965	55	30	1	12	0	2
1973	76	16	2	5	1	0.2
1975	72	17	3	6	2	0.3
1978	73	13	5	5	4	0
1980	65	17	7	6	5	0

Table 1.6 (*continued*)

Energy-Consuming Sector and Year	Petroleum	Coal	Natural Gas	Hydro	Nuclear	Firewood
Korea						
Power generation						
1965	12	70	0	18	0	0
1969	63	21	0	16	0	0
1973	81	11	0	8	0	0
1979	80	4	0	7	9	0
1980	77	8	0	6	10	0
Nongeneration[a]						
1965	12	41	0	0	0	47
1969	39	33	0	0	0	28
1973	50	33	0	0	0	17
1979	59	33	0	0	0	8
1980	57	36	0	0	0	7
Total						
1965	12	44	0	2	0	43
1969	42	32	0	2	0	25
1973	55	30	0	1	0	14
1979	63	27	0	1	2	7
1980	61	30	0	1	2	6
United States						
Power generation						
1965	7	53	22	19	0.3	0
1973	18	43	19	15	5	0
1975	16	43	16	16	9	0
1978	16	44	14	14	13	0
1981	10	51	15	12	12	0
Nongeneration[a]						
1965	53	15	33	0	0	—
1973	57	8	34	0	0	—
1975	59	8	33	0	0	—
1978	63	7	31	0	0	—
1981	60	7	33	0	0	—
Total						
1965	43	23	31	4	0.1	—
1973	47	18	30	4	1	—
1975	46	18	28	5	3	—
1978	49	18	26	4	4	—
1981	43	22	27	4	4	—

a. Includes the industrial, transportation, residential, and commercial sectors.

— Not available.

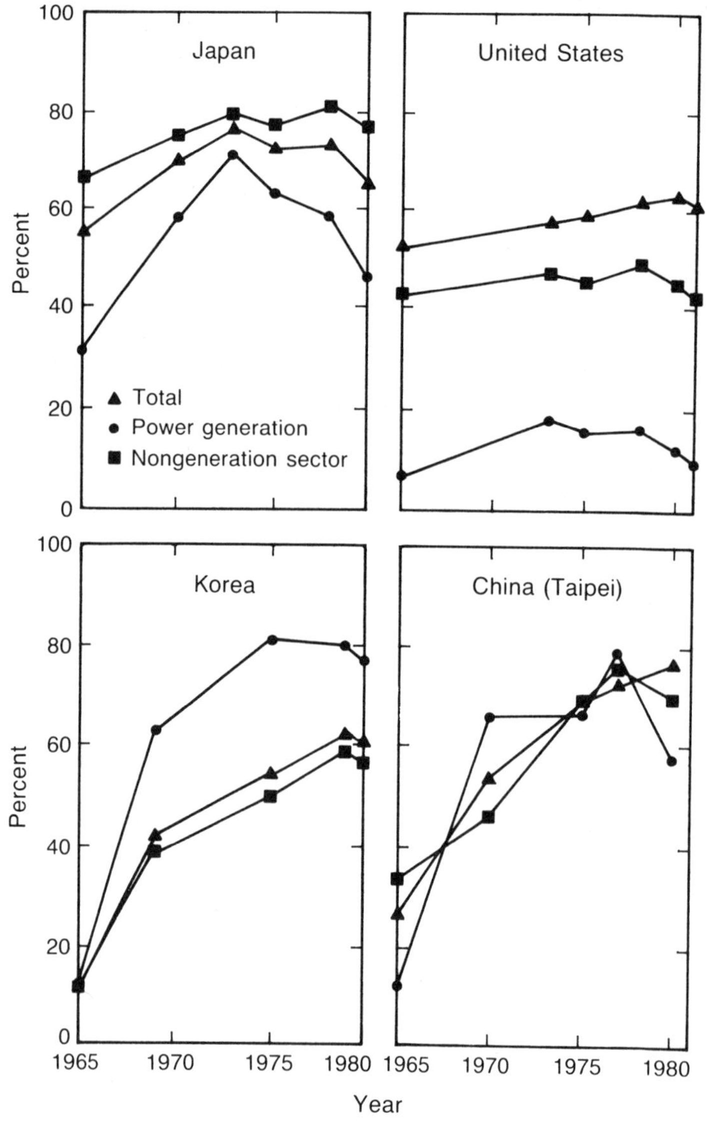

Figure 1.3. Historical sectoral proportions of petroleum as a percentage of primary energy consumption in Northeast Asia and the United States.

CONCLUSION

The electric power sector, perhaps more than any other, has been the battleground for debates about the best government policies to pursue to stimulate economic and social development. This is so for a number of reasons. First, being a natural monopoly, the power sector is always under direct or indirect government control. Thus, it seems more malleable to government policy shifts than other sectors and, as mentioned above, uses a large fraction of national development funds. There is, however, another set of nonfinancial factors related to electricity's image. Because so many of the new activities that are taken up during economic development require electricity, there is a natural tendency to unconsciously perceive electrification to be the *cause* when it may well often be the result of development. For whatever reasons, it is true that to many observers the blessings of electrification have been identified as a prerequisite to and, to some people, as almost a definition of development. This has been true since at least 1920 when Lenin penned his famous and often-quoted sum: Communism equals Soviet government plus electricity.

In the latter part of the period upon which this book focuses (1960–1980), this traditional approach to power development was criticized for several other reasons. Concerns about dependency led to arguments against trends toward large centralized power. It was argued that smaller units and more distributed approaches would provide more reliability, less vulnerability, and greater social and political equity even though traditional cost-benefit analyses tend to show increasing economies of scale.

Another set of arguments came out of the alternative technology movement. Of the various issues addressed, the critics focused on the environmental and resource consequences of heavy reliance on the popular choices of primary fuels to replace oil for power generation (particularly coal and nuclear power). Many of the most promising renewable sources of energy are most usefully applied in making electricity (for example, wind, geothermal, ocean thermal, biomass, and photovoltaic).

A third source of criticism relates to arguments framed in terms of net energy and energy quality. Depending on the type of task being considered and the weights given to cleanliness and convenience at end use, conversion of primary fuels to electricity can be seen to be wasteful. This conclusion stems from the perception that energy resources have intrinsic values beyond what is normally reflected in their relative costs. Thus, it may be important to avoid the "loss" of two-thirds of the energy that typically is dissipated at thermal power plants, even if normal economic considerations favor it.

A more sophisticated and useful framework relates to the idea that fuels should be matched to tasks according to their energy quality. Thus, it may be unnecessarily wasteful to use a high-quality form of energy such as electricity to perform a low-quality task such as heating domestic water. This framework is useful for identifying those tasks in which more energy-efficient equipment might be applied.

These arguments are not entirely separable. For example, they come together in the growing interest in cogeneration, which often can be supported by arguments related to dependency, alternative energy, as well as energy quality. In a similar way, "appropriate" energy technology is often defined by such a combination (Flavin, 1986).

It is clear that some of these arguments are now having important effects on the development of power systems in Northeast Asia. Environmental concerns, in particular, have entered strongly in the same sequence observed for many other changes, first in Japan, then in China (Taipei), and finally in Korea. During the period of primary interest here, however, only Japan was significantly affected. Because of our emphasis on international comparison, we have not addressed these issues in depth. Such a study would, however, likely lead to useful additional lessons for other developing countries.

In the chapters that follow, therefore, we explore the relationship between electricity and economic growth and the evolution of the power industries in some detail for the three Northeast Asian countries. We believe such an exploration will help provide a more objective look at electricity's role in development.

NOTES

1. A typical example would be the United States where electricity use has paralleled the GNP with at least two changes in slope during this century. The first change in slope, from about 0.29 to 0.58 kWh/US$GNP, occurred between 1910 and 1920. The other 2.14 kWh/US$GNP occurred in 1947. There may have been another change in slope toward less electricity intensity at about 1974, but there are too few data points yet to be sure.

2. Most of the data presented in this book comes from a set of reports written as part of the Electric Futures Project at the Resource Systems Institute of the East-West Center. Since they are so frequently the source of data, they are not usually cited separately but are listed in the selected bibliography. The basic data related to utility operations found in these reports have been supplied with the kind cooperation of the utilities in each of the countries.

REFERENCES

Flavin, C. (1986). "Electricity for a Developing World: New Directions." Paper No. 70. Washington, D.C.: Worldwatch Institute.

Kahane, A., and Squitieri, R. (1987). "Electricity Use in Manufacturing." *Annual Review of Energy* 12:223–251.

Pachauri, R. K. (1982). "Electric Power and Economic Development: The Case of India." *Energy Policy* 10(3):189–202.

Searl, M. F. (1986). "The relationship of electricity to aggregate economic growth in industrial economies—some empirical observations." In R. K. Pachauri, ed., *Global Energy Interactions,* pp. 601–620. New Delhi: Allied Publishers.

CHAPTER 2

STRUCTURE, PARAMETERS, AND PREDICTION OF DEMAND

Y. H. Kim, O. S. Yu, and M. Nakamura

The growth in sales of electricity in Northeast Asia during the period of principal interest in this study (1960–1982) was accompanied by some important changes in the structure of demand. In this chapter, particular attention is paid to significant changes in the pattern of electricity consumption, the main influences on the demand for electricity, and the forecasting methods used by the utilities in planning. The chapter also suggests new approaches and innovative concepts useful for forecasting in today's uncertain environment.

STRUCTURAL CHANGES IN ELECTRICITY CONSUMPTION

Analysis of electric power demand (in kilowatts), electric energy demand (in kilowatt hours), and load shape are starting points in a comprehensive economic and technical evaluation of the postpetroleum electric future of China (Taipei), Japan, and Korea. This is true because load shape, peak load, and electricity consumption are the important determinants of future generation capacity additions. Table 2.1 presents historical data and shows that a high growth rate was evident during the period of rapid economic growth from 1955 to 1973. From 1955–1965 to 1965–1973, the growth rate accelerated for China (Taipei) and Korea, whereas it remained virtually the same for Japan. As a consequence of the first oil shock, however, electricity growth in Japan sharply declined to 0.8 percent per annum between 1973 and 1975. On the other hand, electricity demand in China (Taipei) and Korea continued to rise at 8.8 and 16 percent, respectively, although at a substantially lower rate than from 1955 to 1973. More dramatically, electricity consumption dropped in Japan in 1974 for the first time in the postwar period and grew by a scant 3 percent in 1975.

Table 2.1
Growth Rates of GNP, Commercial Energy Use, and Electricity Use, 1955–1981, by Percentage

			Adjustment Period of First Oil Shock				Adjustment Period of Second Oil Shock		
	1955–1965	1965–1973	1974	1975	(1973–1975)	1975–1979	1980	1981	(1979–1981)
China (Taipei)									
GNP[a]	8.1	11.0	1.1	4.2	(2.7)	11.3	6.6	5.5	(5.8)
IIP[b]	11.2	17.4	-6.0	7.7	(0.6)	16.7	6.7	5.6	(6.1)
Energy[c]	8.5	11.7	0.6	10.9	(5.6)	14.2	4.7	-4.6	(-0.1)
Electricity	14.3	15.5	5.3	12.4	(8.8)	13.6	7.4	-1.3	(3.0)
Industrial use	15.1	14.2	5.9	10.2	(8.0)	13.7	7.1	-3.5	(1.7)
Residential/commercial use	12.1	18.8	4.2	17.1	(10.5)	12.9	8.0	3.2	(5.6)
Japan									
GNP[a]	9.2	9.5	-0.2	3.6	(1.7)	5.2	3.7	3.0	(3.2)
IIP[b]	13.6	12.7	-9.7	-4.4	(-7.1)	7.5	4.6	3.6	(4.1)
Energy[c]	11.8	12.0	-2.6	-1.2	(-1.9)	3.5	-4.4	-2.1	(-3.3)
Electricity	12.4	12.1	-1.4	3.0	(0.8)	5.4	-1.7	0.5	(-0.6)
Industrial use	12.1	11.5	-3.5	-0.4	(-1.9)	4.3	-2.5	-0.6	(-1.5)
Residential/commercial use	14.5	14.7	4.0	11.6	(7.8)	8.2	-0.1	4.4	(2.1)
Korea									
GNP[a]	5.0	10.2	8.0	7.1	(7.6)	10.8	-6.2	6.4	(0.0)
IIP[b]	—	20.8	27.8	19.1	(23.4)	21.0	-1.9	10.5	(4.1)
Energy[c]	15.3	15.5	2.8	7.4	(5.0)	13.8	2.5	4.3	(3.4)
Electricity	16.2	22.3	13.6	18.4	(16.0)	17.0	5.1	8.2	(6.6)
Industrial use	—	24.0	16.0	18.7	(17.4)	16.4	5.2	6.0	(5.7)
Residential/commercial use	—	19.1	8.1	17.5	(12.7)	18.4	4.4	13.3	(8.8)

Sources: Ke et al. (1983); Kim et al. (1983); Kimura et al. (1983).
a. In real terms. b. Index of industrial production. c. Commercial enegy. — Not available.

As the economy recovered from the first oil crisis, however, electricity demand in the three countries resumed a rapid rise in the 1975–1979 period until the second oil shock (see Table 2.1). During this period in China (Taipei) and Korea, the annual average growth rate of the Gross National Product (GNP) in real terms rose at an even higher rate than before, and even though electricity consumption rose rapidly during this period, the rate of increase was slightly less than during the pre-embargo period 1965–1973. In contrast, during 1975–1979, after the adjustment period of the first oil shock, the Japanese economy and electricity consumption grew at rates lower than half of annual growth rates in the period 1965–1973.

In 1979–1980, a doubling of petroleum prices resulted in another major perturbation of electricity demand. Even though the annual average growth rate of the GNP in China (Taipei) and Japan rose substantially higher from 1979 to 1981 than during the 1973–1975 period, as Table 2.2 indicates, the annual growth rate of electricity consumption fell markedly during the adjustment period of the second oil shock compared with the 1973–1975 period. On the other hand, the Korean economy experienced a negative growth rate of GNP in 1980 for the first time in the postwar period, but electricity consumption grew at 5.1 percent. Japan in 1980 and China (Taipei) in 1981 experienced a negative growth rate of electricity consumption even though real growth of the GNP in both countries was positive.

There are also wide variations in the structural changes of electricity consumption in the three countries, related to a considerable extent to the degree of industrialization. As Table 2.2 shows, Japanese GNP elasticity fell to 0.47 during the 1973–1975 period and more sharply to 0.17 during the 1979–1981 period. In China (Taipei) the demand for electricity was not much affected by the first oil crisis but had decreased by the second oil crisis. Unlike its two neighbors, on the other hand, Korean electricity consumption, in relation to the GNP, was scarcely affected by both oil crises (see Table 2.2). The trend of high annual growth rates of electricity consumption in Korea was, however, substantially dampened by the second oil crisis (see Table 2.1).

In general, the impact of the second oil crisis on the demand for electricity turned out to be stronger than that of the first oil crisis. This can be attributed partly to differences in perception by the public, who regarded the first oil crisis as a one-shot affair and the second, because of the increases in OPEC (Organization of Petroleum Exporting Countries) prices, as a continuous, long-run phenomenon. It also can be attributed partly to the inflexibility in the timely modification of existing capital equipment from existing production methods to methods appropriate to higher energy prices. As a result, there was a sharp turn toward conservation in a broad sense. Thus the second oil shock brought about basic changes in the structure of electricity consumption.

A close look at Table 2.1 reveals significant changes in the pattern of electricity consumption in the three countries of Northeast Asia after the two

Table 2.2
Income Elasticities of Energy and Electricity Use, 1955–1981

			Adjustment Period of First Oil Shock		Adjustment Period of Second Oil Shock
	1955–1965	1965–1973	1973–1975	1975–1979	1979–1981
China (Taipei)					
Energy[a]	1.05	1.06	(2.07)	1.26	—
Electricity	1.77	1.41	(3.26)	1.20	(0.51)
industrial use	1.35	0.83	(13.70)	0.81	(0.26)
residential/commercial use	1.49	1.71	(3.88)	1.14	(0.97)
Japan					
Energy[a]	1.20	1.23	—	0.67	—
Electricity	1.35	1.27	(0.47)	1.04	(.17)[b]
industrial use	0.89	0.91	—	0.57	—
residential/commercial use	1.58	1.55	(9.75)	1.58	(0.66)
Korea					
Energy[a]	3.06	1.52	(0.66)	1.27	(0.67)[b]
Electricity	3.24	2.19	(2.11)	1.57	(1.28)[b]
industrial use	—	1.15	(0.75)	0.78	(0.57)[b]
residential/commercial use	—	1.87	(1.69)	1.70	(1.38)[b]

Sources: Ke et al. (1983); Kim et al. (1983); Kimura et al. (1983).

Note: Elasticities of total energy, electricity use, and residential use are measured with respect to increases in GNP, and for industrial power use with respect to increases in output.

a. Commercial energy. b. 1981 figures. — Not available.

oil shocks. The annual average growth rate of electricity use fell more sharply from 1979 to 1981 than from 1973 to 1975. On the other hand, in China (Taipei) and Japan the annual growth rate of the GNP declined more sharply during the first oil shock period than during the second. Consequently, as shown in Table 2.2, the GNP elasticity of electricity consumption declined from 1.20 during the recovery period 1975–1979 to 0.51 during the 1979–1981 period in China (Taipei), from 1.04 to 0.17 in Japan, and from 1.57 to 1.28 in Korea. It has since then remained at the lower level.

The second basic change the three countries have undergone in their electricity consumption pattern, though in different degrees, is the unbalanced growth of electricity consumption by end-use sectors. As Table 2.1 shows, the annual growth rate of residential and commercial electricity consumption outstripped that of industrial use from the 1955–1965 period onward in Japan, from 1965 to 1973 in China (Taipei), and only from 1975 to 1979 in Korea. Moreover, in the three countries the adverse impact of the two oil shocks turned out to be more severe for industrial users than it did for residential users. Industrial power consumption in Japan dropped not only during the second oil crisis period, but also during the first, whereas industrial use in China (Taipei) remained relatively unaffected by the first oil crisis but recorded a negative growth rate in 1981. On the other hand, in Korea the growth rate fell from 16.0 percent to 5.7 percent between 1975–1979 and 1979–1981 (see Table 2.1). Consequently, as Table 2.2 shows, the elasticity of industrial power use with respect to the increase of industrial output sharply declined from 0.57 to a minus value between 1975–1979 and 1979–1981 in Japan, and from 0.81 to 0.26 in China (Taipei).

As a result of the sharp rise in energy prices, the industry chose a mix of production factors with less energy. More importantly, the Japanese economy has undergone drastic structural changes, shifting away from energy-intensive materials in industry. In Japan, since the oil shock in 1973, economic growth has become more dependent on tertiary industry, which mainly consists of the services and the high value-added machinery sectors, rather than on the growth of energy- and electricity-intensive materials production, such as steel and chemical raw materials (see Figures 2.1 and 2.2). With the second oil shock as a turning point, the comparative advantage of China (Taipei) and Korea in international trade has also shifted against the petrochemicals and nonferrous metals industries and some steel-making technologies (direct reduction/electric arc furnaces). As Table 2.3 shows, the proportion of power-intensive industrial power demand in total manufacturing use declined between 1980 and 1981 in the three countries, more dramatically in Japan, next in China (Taipei), and last in Korea.

In contrast to the adverse impact of the oil shocks on industrial use, residential and commercial use of electricity was scarcely checked by the first oil shock in the three countries of Northeast Asia. During the adjustment period of the second oil crisis (1979–1981), however, the annual average growth of residential and commercial electricity consumption substantially declined compared with the previous period, but still grew at a positive rate.[1] As a result, the growth of electricity demand from 1955 to 1980 was accompanied by some important changes in the proportion of total electricity consumption accounted for by the principal classes of consumers (see Table 2.4). From 1955

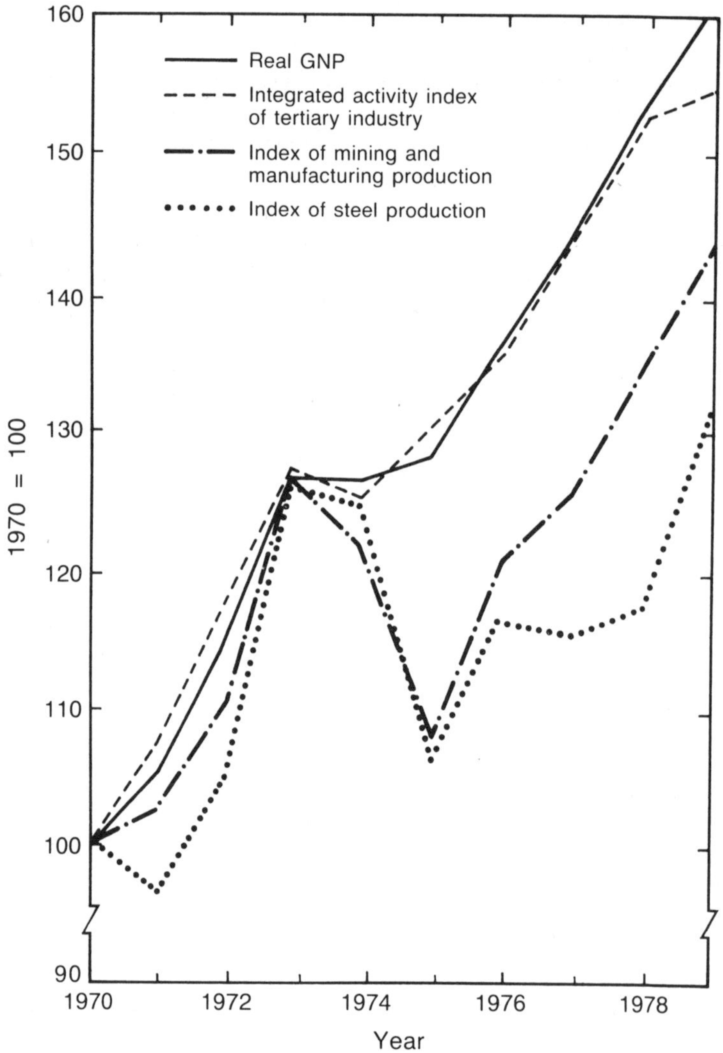

Figure 2.1. Historical change of Japanese real GNP with production in mining and manufacturing.

to 1962, sales to industrial consumers rose from 70 to 77 percent in China (Taipei), but declined thereafter. In Japan the proportion of total sales accounted for by industrial consumers rose from 76 to 79 percent between 1955 and 1960, but declined thereafter. The industrial proportion in Korea rose from 62 percent in 1965 to 73 percent in 1976 and declined thereafter. Thus with the years 1960 in Japan, 1962 in China (Taipei), and 1976 in Korea as turning points, residential and commercial sales rose more rapidly than industrial sales and, consequently, the relative importance of residential and commercial sales became more apparent. The proportions of industrial sales reported in 1980 in the three countries are, nonetheless, still very high, at 60 to 70 percent, when compared to the 39 percent in the United States in recent years. The industrial share of total electricity demand has a strong correlation with the real per capita GNP in the industrialized countries, but the share tends to decline steadily as the per capita GNP rises (see Figure 2.3).

The third fundamental change in consumption patterns in the three countries is the rapid growth of peak load in contrast to the stagnating growth of baseload. This is mainly caused by the rapid increase in cooling demand, which is discussed in Chapter 3. Living standards improved with rapid economic growth, and the use of air conditioners and refrigerating electric appliances increased accordingly.

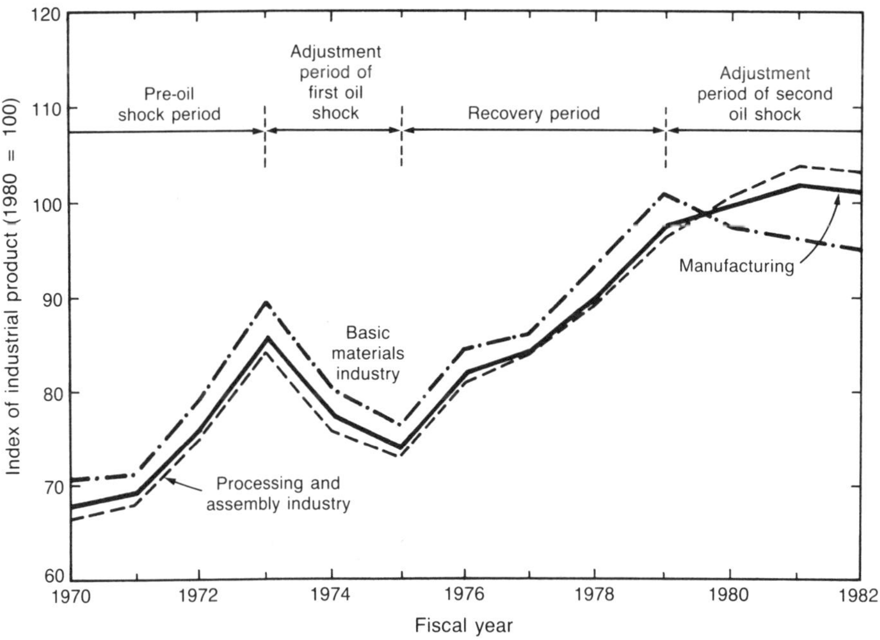

Figure 2.2. Historical trends of Japanese industrial production.

Table 2.3
Composition of Manufacturing Power Consumption by Major Sectors, 1970–1981, by Percentage

	1970	1975	1979	1980	1981
China (Taipei)					
Light manufactures	30.3	37.3	—	32.9	34.6
Textiles	(13.3)	(18.5)		(15.5)	
Power-intensive manufactures	62.4	53.4	—	55.9	53.3
Chemicals	(28.4)	(27.3)		(25.0)	—
Basic metals	(17.5)	(12.3)		(15.7)	—
Technology-intensive manufactures	7.3	9.3	—	11.1	12.1
Japan					
Light manufactures	7.3	7.5	9.8	9.9	10.6
Power-intensive manufactures	86.5	84.5	80.5	79.5	77.7
Chemicals	(32.1)	(26.8)	(19.9)	(18.4)	(18.5)
Iron and steel	(36.1)	(40.1)	(41.2)	(41.2)	(39.3)
Technology-intensive manufactures	6.2	8.0	9.7	10.6	11.7
Korea					
Light manufactures	30.1	32.3	31.9	30.1	29.9
Textiles	(16.4)	(21.1)	(21.1)	(20.7)	(20.8)
Power-intensive manufactures	66.3	59.2	57.0	59.4	58.7
Chemicals	(27.2)	(21.9)	(18.4)	(19.4)	(19.4)
Basic metals	(16.9)	(16.8)	(19.9)	(21.2)	(21.2)
Technology-intensive manufactures	3.6	8.5	11.1	10.5	11.4

Sources: Ke et al. (1983); Kim et al. (1983); Kimura et al. (1983).

— Not available.

As in other countries, a dominating summer peak has been associated with rapid growth of daytime cooling load. In particular, in Japan the proportion of cooling demand in the peak demand in August increased rapidly from 17 percent in 1968 to 31 percent in 1978 (see Figure 2.4). Whereas China (Taipei) and Korea are approximately fifteen and twenty years, respectively, behind Japan in the level of per capita GNP, they lag behind Japan in cooling demand by only four and thirteen years, respectively. The main factors explaining why cooling demand patterns are different at the same level of per capita GNP in the three countries are the climate and the demonstration effect. Geographically located in a subtropical region, the GNP elasticity of cooling demand is the highest in China (Taipei). In 1984 air conditioners used more than 22 percent of residential electricity consumption compared to less than 8 percent in 1975. Figure 2.5 shows dramatically the effect of weather as measured in cooling-degree-days on peak load.

In 1981 the diffusion rate of air conditioners differed among Japanese cities, ranging from 3.9 percent in Hokkaido to 60 percent in Tokyo and to 90 percent in Osaka—cities where the average temperatures in August 1981 were

Table 2.4
Composition of Sectoral Power Demand, 1955–1981, by Percentage

	Residential and Commercial	Industrial	Transportation
China (Taipei)			
1955	30.1	69.6	0.3
1962	22.7	77.1	0.2
1972	30.5	69.3	0.2
1981	33.2	65.9	0.9
Japan			
1955	17.0	76.0	7.0
1960	16.0	79.0	5.0
1968	22.0	74.0	5.0
1980	33.0	64.0	3.0
Korea			
1955	—	—	—
1965	38.0	62.0	a
1976	27.0	73.0	a
1981	31.0	69.0	a

Sources: Ke et al. (1983); Kim et al. (1983); Kimura et al. (1983).

a. Included in industrial use.

— Not available.

17, 23, and 26 degrees centigrade, respectively. This demonstrates the importance of the climate factor for cooling demand.

Since the diffusion rates of air conditioners in 1981 ranged from 56 percent in Japan to 17 percent in China (Taipei) and 4 percent in Korea, there is a potential for further increases in daytime cooling demand during the coming years, thereby affecting the annual load factors in the three countries (see Chapter 3). This potential becomes more apparent from the Chinese and Japanese experiences in the evolution of diffusion rates of major home electric appliances for the past twenty years shown in Figures 2.6 and 2.7. The

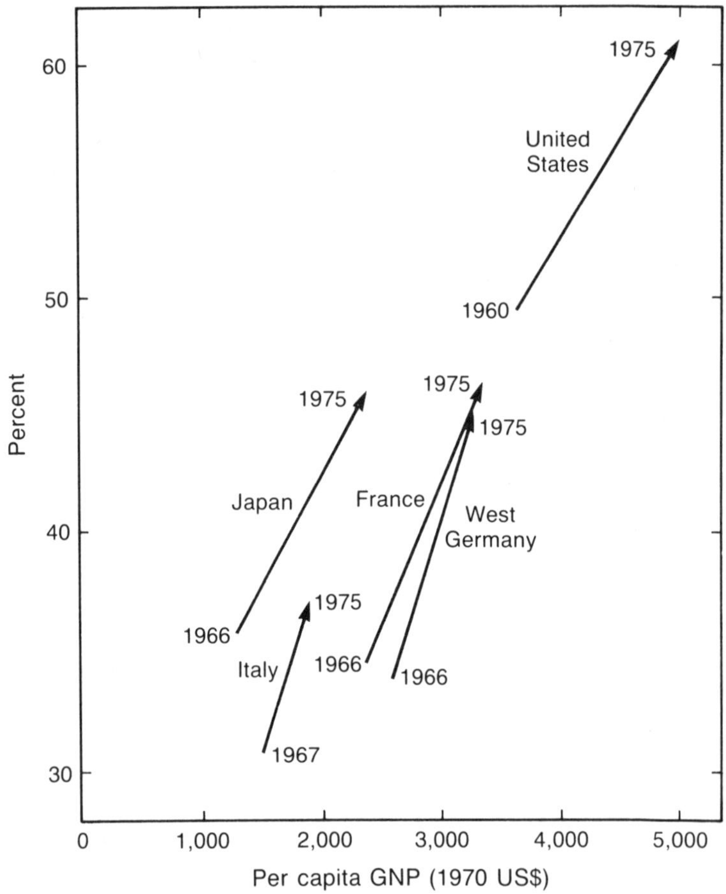

Figure 2.3. International comparison of the nonindustrial percentage of total electricity consumption. Note the sharp rise in all industrialized countries during the 1960s and early 1970s.

diffusion of major appliances in China (Taipei) has been fairly uniformly distributed in rural and urban areas, a measure of the rural/urban income equity the country enjoys.

Finally, as reflected in the upward trend of the electricity share in total primary energy consumption (i.e., the share of primary energy consumed in the form of electricity) in Table 2.5, Japan has been undergoing a transition from oil-based economy to electricity-based economy, whereas China (Taipei) and Korea are already following suit. This so-called electricity shift can be attributed to the relative lower price of electricity compared with fossil fuels, as seen in Table 2.6, and the nature of the industrialization process. Table 2.1 shows that the annual growth rate of energy use in Japan was more severely depressed by the two oil shocks, compared with the annual growth rate of electricity use. The rapidly progressing electricity shift, particularly after the oil shock, has become a common phenomenon even in the United States and Western Europe. On the other hand, in China (Taipei) and Korea it was only during the second oil shock that the annual growth of energy consumption declined more rapidly than that of electricity consumption.

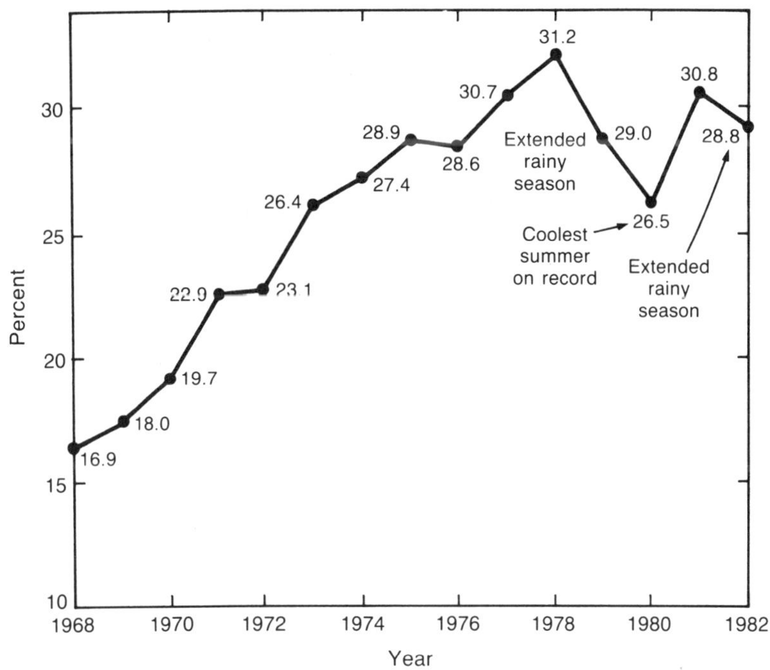

Figure 2.4. Historical change in the cooling demand (air-conditioning and refrigeration) percentage of the summer peak in Japan.

MAJOR FACTORS DETERMINING ELECTRIC POWER DEMAND

In the last two decades electricity consumption has grown at average annual rates of 13 percent in China (Taipei), 8.6 percent in Japan, and 19 percent in Korea. This rapid growth stemmed from persistent increases in demand in all major sectors. Theoretical considerations suggest that the consumption of electricity is primarily determined by economic variables, demographic

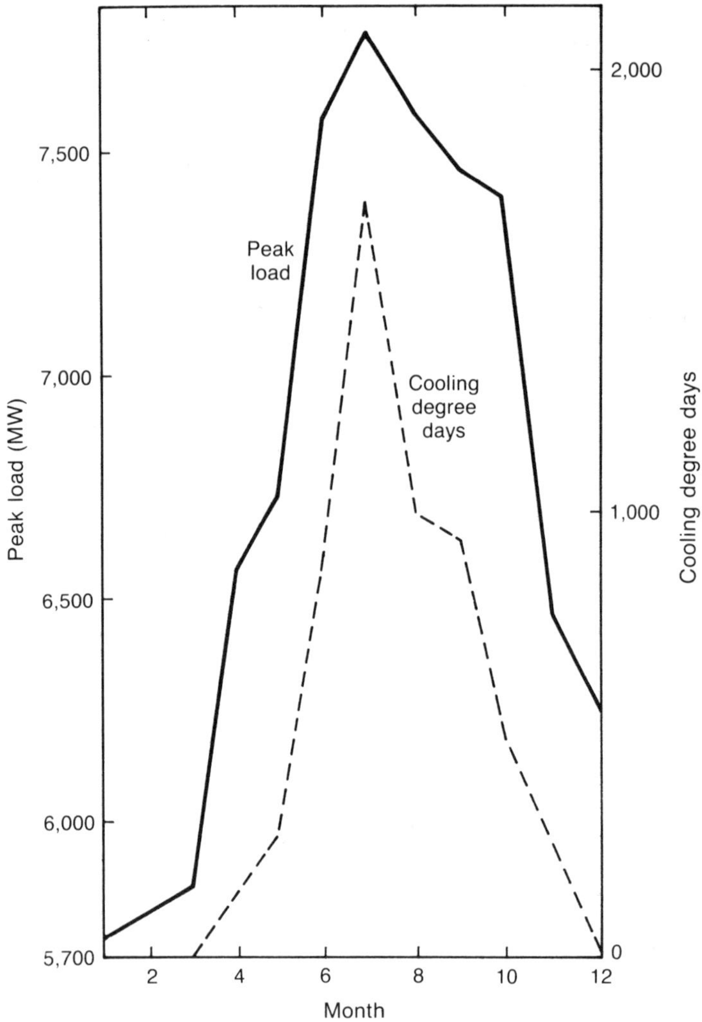

Figure 2.5. Relationship of peak load and cooling degree days during 1983 in China (Taipei).

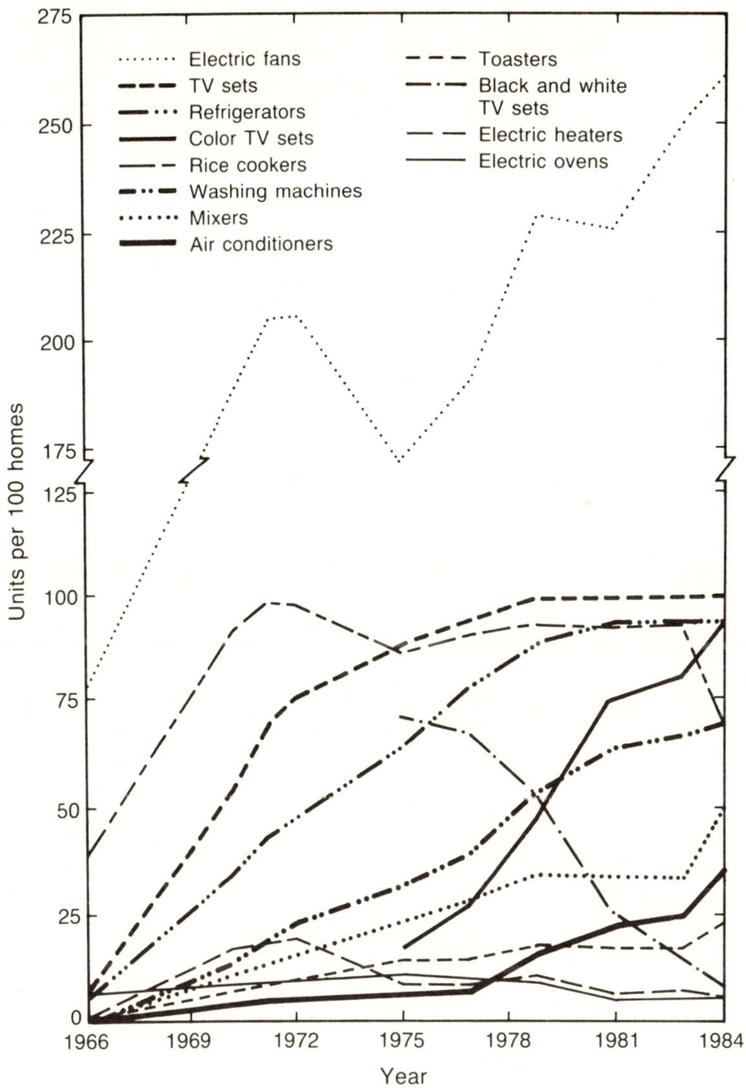

Figure 2.6. Historical penetration of major home appliances in China (Taipei).

variables, and the impact of conservation (see Table 2.7). In considering these factors, special attention is paid to sales to residential and industrial consumers, who together accounted for more than 85 percent of total sales in the three countries in most years. As noted in Table 2.1, from 1965 to 1980, sales for residential purposes were the fastest growing element of demand for electricity.

Major determinants of this demand have been the growth in the number of households served by electricity and increases in the level of real income per household, which generally leads to greater expenditure on household appliances (see Table 2.8). In China (Taipei) the proportion of electricity consumption by household appliances rose rapidly from 50 percent in 1975 to 61 percent in 1983. Relative prices of electricity and alternative forms of energy

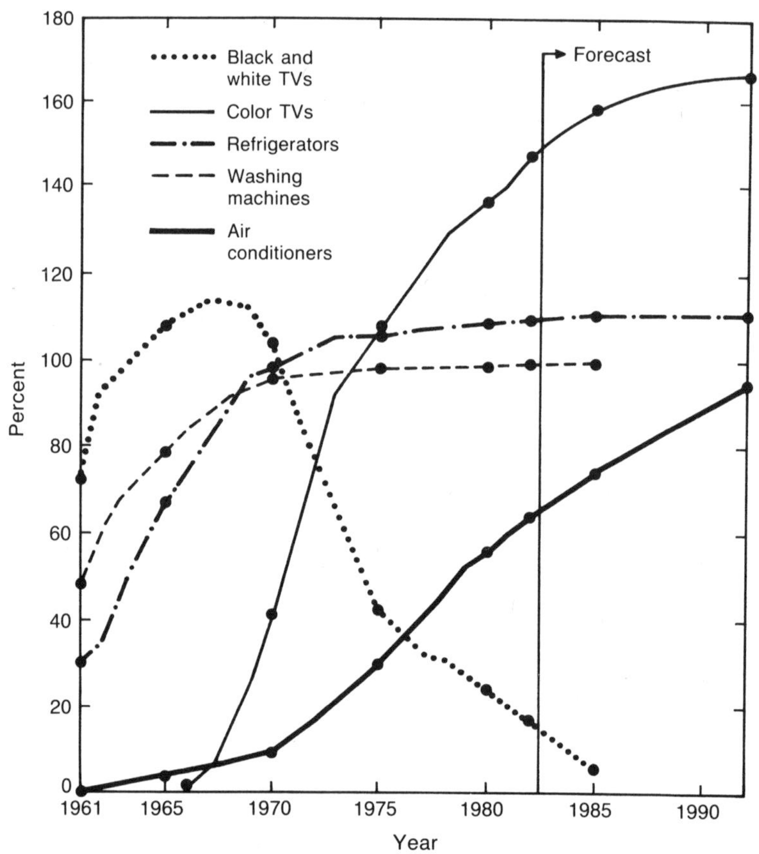

Figure 2.7. Historical and forecasted penetration of major home appliances in Japan.

can also be expected to play a role in determining levels of household consumption of electricity in the long run mainly by influencing the type of appliance purchased by consumers.

Between 1960 and 1980 the number of households served by electricity suppliers increased at a much faster rate than did population, the average annual rate ranging from 4.5 percent in Japan to 7.4 percent in China (Taipei) and 10.6 percent in Korea (see Table 2.8). A high rate of growth in residential demand has also been fueled by the rapid urbanization of this period, the urbanization rate sharply rising from 50 to 70 percent in China (Taipei), from 64 to 76 percent in Japan, and from 28 to 67 percent in Korea. More fundamental is the phenomenal increase in real income per capita. As shown in Table 2.8, between 1960 and 1980 per capita real GDP (Gross Domestic Product) rose greatly at average annual rates of 7.1 percent in China (Taipei), 6.4 percent in Japan, and 6.7 percent in Korea. It is apparent in all three countries that the proportion of residential energy consumption in total energy consumption rose with economic growth. At the same time the electricity share in total household energy consumption also apparently rose (see Figure 2.8).

Whereas the real income per capita of China (Taipei) increased most rapidly among the three countries during 1960–1980, the fastest growth of electricity consumption occurred in Korea. This can be explained by Korea having both the lowest initial level of per capita electricity consumption and the highest rate of growth in the number of consumers among the three countries. The outcome is also heavily influenced by the colder climate of Korea and the pattern of industrial uses in all three countries. Table 2.8 shows that Korea's per capita electricity consumption was 62 kWh at the start of the period, China's (Taipei), 312 kWh, and Japan's, 1,064 kWh.

Among other factors contributing strongly to higher levels of consumption was the rural electrification of Korea. In 1964 fewer than 12 percent of all rural dwellings reported the availability of electric power, but by 1977, 100 percent of all rural dwellings had been provided with electricity. The fall in

Table 2.5
Electricity Consumption as a Percentage of Total Energy Consumption, 1965–1981

	China (Taipei)	Japan	Korea
1965	30.7	28.8	14.7
1975	33.9	31.1	20.8
1980	34.7	35.1	21.8
1981	—	36.4	—

Sources: Ke et al. (1983); Kim et al. (1983); Kimura et al. (1983).

— Not available.

Table 2.6
Real Energy Prices of Major Fuels in Japan (1970 = 100)

Use/Type of Fuel	1970	1973	1976	1979
Residential and commercial uses				
Kerosene	100	89	110	117
LPG	100	98	100	95
City gas	100	92	105	104
Electricity	100	79	77	72
Industrial uses				
Bunker-C	100	110	218	209
Coal	100	90	156	130
Electricity	100	92	151	153

Sources: Ke et al. (1983); Kim et al. (1983); Kimura et al. (1983).

the price of electricity relative to competing fuels and the fall of home electric appliance prices also appear to have contributed to the increase in consumption of electricity in those countries. The real price of electricity in 1980 was more than 32 percent above the 1975 level in China (Taipei), 46 percent in Japan, and 48 percent in Korea. It can be postulated that to a great extent the increase in the price of electricity at a rate higher than inflation has sparked conservation.

Conservation behavior has been reinforced by the introduction of increasing block rate, peak load pricing, and electricity-saving technology. In China (Taipei), sales to commercial consumers declined by 3.4 percent after higher prices for users of more power were put into effect in 1979. Present Korean experience indicates a 10 percent reduction in peak load since the implementation of time-of-use rates (see Chapter 3). Some studies in Japan also report that end-use efficiency for new electric appliances improved considerably between 1973 and 1980. These factors may help explain in part the fall in GDP elasticity of electricity consumption (see Table 2.9).

During the 1975–1980 period, however, the real price rise of petroleum far exceeded that of electricity in China (Taipei), Korea, and Japan. As a result, it is difficult to determine the impact of conservation and substitution effects on the consumption of electricity. More significantly, the drastic fall in the real price of household electric appliances in the 1975–1980 period may have influenced consumers to purchase electric appliances and may have continued to deter consumers from switching to appliances using other fuels, even when there was some real increase in the price of electricity. It is particularly noteworthy that the real price of home electric appliances in Korea in 1980 sharply declined to 40 percent of the 1975 level.

Table 2.7
International Comparison of Residential Electricity Consumption per Household by Major Electric Appliances in 1977

Item	Japan			England			United States		
	Average Electricity Consumption (kWh/yr)	% of Total Electricity Consumption per Household	Diffusion Rate (%)	Average Electricity Consumption (kWh/yr)	% of Total Electricity Consumption per Household	Diffusion Rate (%)	Average Electricity Consumption (kWh/yr)	% of Total Electricity Consumption per Household	Diffusion Rate (%)
Refrigerator	584	24	105.0	475	11	107	2,071	33.0	169
Television	369	15	154.0	358	9	98	308	4.9	157
Lighting	404	17	—	329	8	—	900	14.3	—
Electric cooker or range	90	4	5.4	417	10	42	420	6.7	60
Floor heater and room cooler	288	12	72.0	846	20	81	1,104	17.5	38
Water heater	256	11	6.0	1,090	26	68	392	6.2	12
Other appliances	405	17	—	652	16	—	1,086	17.4	—
Total electricity consumption per household (kWh)	2,391	100		4,167	100		6,281	100.0	

Sources: Ke et al. (1983); Kim et al. (1983); Kimura et al. (1983).

Note: The United States is represented by the San Francisco area, whose service is provided by the Pacific Gas and Electric Company. Because the climate in San Francisco is mild, electricity consumption for room heating and cooling is relatively small. Moreover, abundant natural gas is widely used for water heating and cooking. It should be understood that the entire U.S. average electricity consumption is greater for these uses.

— Not available.

Table 2.8
Factors Determining Power Demand, 1960–1980

	China (Taipei)	Japan	Korea
(1) Annual growth rate of power demand (%)	13.8 (10.3)[a]	8.7 (7.4)	19.3 (14.0)
Demographic factors			
(2) Population growth rate (%)	2.5	1.1	2.1
(3) Growth rate of households serviced by electricity supplies (%)	7.4	4.5	10.6
(4) Urbanization rate (%)			
1960	50	64	28
1980	70	76	—
1981	—	—	67
Economic factors			
(5) Initial level of power demand (kWh/person)	312	1,064	62
(6) Annual growth rate of real GDP (%)	9.6	7.6	8.6
(7) Annual growth rate of per capita real GNP (%)	7.1	6.4	6.7
(8) Industrial structure (output share of electricity-intensive manufacturers) (%)			
1960	—	—	14
1965	—	33	—
1970	26	37	36
1975	36	37	36
1978	—	—	37
1979	—	36	—
(9) Technological advances (share of electricity in industrial demand for energy) (%)			
1965	33	33	18
1980	34	37	31
(10) Rural electrification (%)			
1964	84	—	12
1977	100	—	100
(11) Real price indices in 1980 (1975 = 100)			
electricity	127	146	148
oil	151	168	156
electric appliances	76	75	40

Table 2.8 (*continued*)

	China (Taipei)	Japan	Korea
Conservation			
(12) Tariff rate and load management techniques	reverse block rate (residential)	reverse block rate (time-(of-use)	reverse block rate (time-of-use)
(13) Electricity-saving technology (index of electricity consumption per unit in 1980; 1973 = 100)			
refrigerator	—	45	—
color television set	—	61	—
air conditioner	—	70	—
vacuum cleaner	—	84	—

Sources: Ke et al. (1983); Kim et al. (1983); Kimura et al. (1983).

a. Figures in parentheses are annual growth rates of per capita power consumption.

— Not available.

For statistical analysis, the following regression equation was estimated for China (Taipei), Japan, and Korea:

$$\text{Ln ER/C} = a + b \text{ Ln DI/N} + c \text{ Ln PE} + d \text{ Ln PF} \qquad (2.1)$$

where ER/C is average residential sales per customer, DI/N is personal disposable real income per capita, PE is an index of the average unit price of electricity in residential use deflated by the consumer price index, and PF is an index of the average unit price of fuel in residential use deflated by the consumer price index.

As item (1) of Table 2.8 shows, real income and price variables, combined with the number of households using electricity, turned out to be significant determinants of residential demand for electricity. When relative growths of residential electricity consumption per household using electricity and personal disposable real income per capita are compared, the implied income elasticity of residential consumption of the 1965–1981 period was approximately 0.86 in Japan, 0.97 in China (Taipei), and 1.27 in Korea.[2] Different income elasticities obviously reflect differences in the living standard and accordingly the degree of saturation of electric home appliances in the three countries. The statistical results from the regression analysis also appear to conform to the results from other research that indicate a considerable degree of responsiveness to prices in residential demand for electricity (Houthakker et al., 1974; Hawkins, 1975).

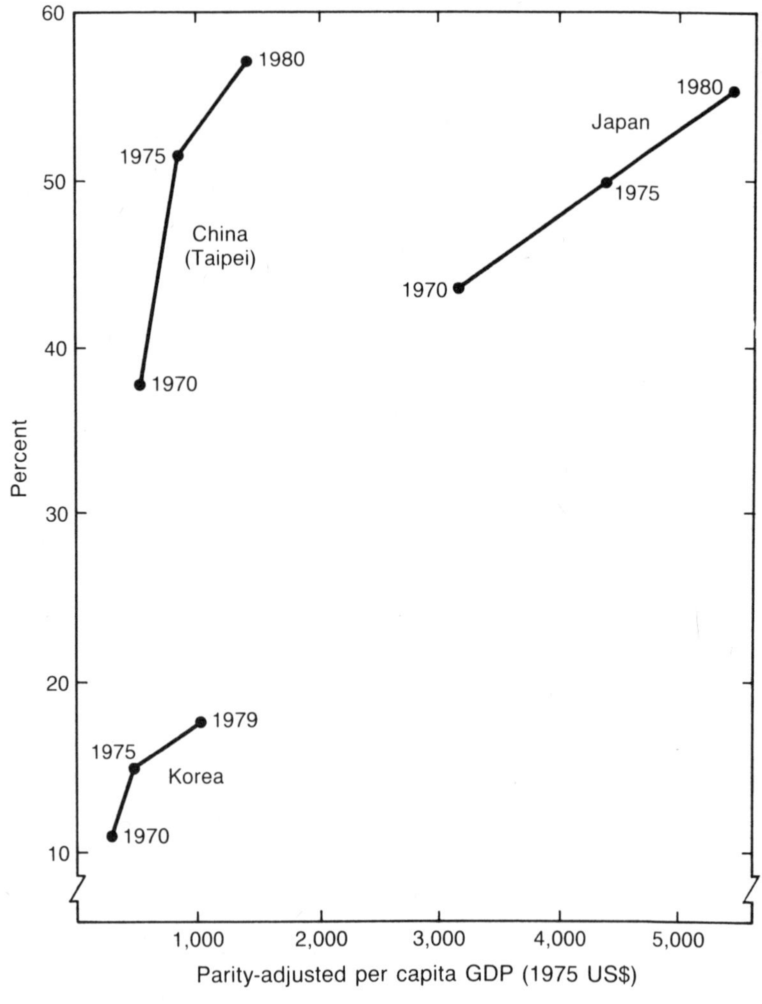

Figure 2.8. Residential electricity consumption as a percentage of total residential and commercial energy consumption in Northeast Asia. Note that the rise during the 1970s was much more rapid in Korea and China (Taipei) than in Japan.

In all equations the signs of the price variables are economically meaningful. The higher price effect in Japan is largely related to the impact of *energy-saving technological* progress in household electrical appliances after the oil shock (see item 13 of Table 2.8). More significantly, the price of competing fuels was a stronger influence on the increase in residential sales in Korea than was the price of electricity paid by residential consumers. Thus the substitution effect dominated the conservation effect in the demand for residential use of electricity in Korea.

In an attempt to establish the relationship between industrial consumption of electricity as the dependent variable and the volume of industrial production and the real prices of electricity and competing fuels as indepen-

Table 2.9
International Comparison of Income and Price Elasticities of Electricity Demand, 1965–1981

	Elasticity Coefficients				
Type of Demand/Country	Income/ Output	Electricity Price	Fuel Price	R^2	D-W
(1) Residential demand					
China (Taipei)	0.97	-0.16	—	0.95	0.79
	(10.40)[a]	(-1.10)			
Japan	0.86	-0.51	0.66	0.99	1.04
	(3.60)	(2.60)	(3.80)		
Korea	1.27	-0.35	0.54	0.99	1.34
	(15.70)	(-2.10)	(3.40)		
(2) Industrial demand					
China (Taipei)	0.96	-0.33	0.23	0.97	1.09
	(8.80)	(0.40)	(1.30)		
Japan	0.79	-0.06	0.06	0.99	1.26
	(8.80)	(-0.30)	(0.30)		
Korea	1.10	-0.01	—	0.99	0.64
	(39.50)	(0.01)			
(3) Iron and steel					
Japan	0.89	-0.26	—	0.97	1.38
	(17.30)	(-2.80)			
Korea	0.89	-0.25	—	0.97	1.80
	(11.30)	(-0.46)			

Sources: Ke et al. (1983); Kim et al. (1983); Kimura et al. (1983).
a. Figures in parentheses are t-values for the estimated coefficients.
— Not available.
D-W Durbin-Watson Statistics.

dent variables, the following regression equations were estimated using the time series data from China (Taipei), Japan, and Korea:

$$\text{Ln EM} = a + b \text{ Ln IIP} + c \text{ Ln PE} + d \text{ Ln PF} \qquad (2.2)$$

where EM is total sales to manufacturing industries, IIP is the index of industrial production, and PE and PF are the average prices of electricity and fuel to industrial consumers, respectively, deflated by the wholesale price index.

As item (2) of Table 2.9 indicates, the output elasticity of industrial demand for electricity turns out to be 0.79 in Japan, 0.96 in China (Taipei), and 1.1 in Korea, reflecting differences in the degree of industrialization among the three countries.

The regression equations appear to reveal that a substantial increase in the use of electricity by industrial consumers is attributed to the growth of industrial output. It is clear that a high rate of growth in the GDP (Table 2.8) was the most important determinant of the increase in industrial sales in those countries.

Although the sign of the price variable in the regression equation (2.2) is economically meaningful in the case of Japan and Korea, the price variable turns out to be statistically insignificant and the influence of price changes on the industrial demand for electricity appears to be negligible (see Table 2.9). It has generally proved extremely difficult to show the influence of price changes on the industrial demand for electricity (Taylor, 1975). A possible explanation may be that the proportion of electricity bills in the total production cost of manufacturing industries, for example in Korea, was on the average 2.6 percent in 1980 (see Table 2.10). This relatively low proportion may help explain the price inelasticity of industrial demand for electricity. As item (3) of Table 2.9 indicates, however, regressions for power-intensive industries such as steel and metal reveal a considerable degree of price responsiveness in electricity use. On the other hand, regression analyses of other power-intensive materials such as chemicals and cement do not prove the significance

Table 2.10
Electricity Value as a Percentage of Total Production Cost in Korea, 1980

Industry	Percentage Share
Agriculture and fisheries	0.2
Mining	6.4
Manufacturing	2.6
Transportation	1.6
Wholesale and retail trade	0.9
Services	2.1

Sources: Ke et al. (1983); Kim et al. (1983); Kimura et al. (1983).

of price variables. Another possible explanation may lie in the fact that, unlike home electric appliances, the economic life of industrial electric equipment is so long and the equipment is so capital intensive that competition from other fuels becomes unimportant. Unless the real price of electricity persistently rises for a long period, industrial consumers have little incentive to switch to equipment using other fuels.

In reality, however, the real price of industrial power use increased 7 percent per year between 1973 and 1981 in Korea, compared with 22 percent for fuel oil. It appears that changes in industrial structure and technological advances have been more important in determining industrial demand in the three countries of Northeast Asia (Fisher and Kaysen, 1962; Baxter and Rees, 1968). In actual production activity, it is likely that the choice of the electricity-intensive technology would turn out to be a good policy in order to achieve fuel (petroleum/coal) savings, which are considered to have greater overall savings on energy cost savings.

The ninth factor in Table 2.8 shows a substantial increase in the use of electricity as a source of industrial energy in China (Taipei), Japan, and Korea between 1965 and 1980. According to these data the energy content (heat equivalent) of electricity used in factories as a proportion of the energy content of all fuel and electricity used increased by approximately 13 percentage points in Korea, 4 percentage points in Japan, and 1 percentage point in China (Taipei). Changes that took place between 1965 and 1980 were fairly strongly biased toward greater use of electricity in industry.

This points out that in industrial usage, electricity has replaced many flame-based technologies, with resulting improvements in product quality in those countries. Even so, it is apparent from Table 2.11 that the rise in the share

Table 2.11
Electricity and Fuel as Percentage of Total Energy Consumption in Industry, 1965–1980

	Japan		Korea			China (Taipei)	
	1965	1980	1965	1973	1980	1965	1980
Electricity	33.1	36.6	18.3	24.1	30.5	32.8	34.2
Fuel							
Coal	23.3	21.4	37.1	11.4	17.2	46.7	12.0
Petroleum	42.1	40.9	19.3	60.2	51.9	15.1	46.7
Natural gas	1.5	1.1	0.0	0.0	0.0	5.4	7.1
Firewood	—	—	25.3	4.3	0.4	—	—
Subtotal	(66.9)	(63.4)	(81.7)	(75.9)	(69.5)	(67.2)	(65.8)
Total energy	100.0	100.0	100.0	100.0	100.0	100.0	100.0

Sources: Ke et al. (1983); Kim et al. (1983); Kimura et al. (1983).

— Not available.

of electricity was accompanied by an increase in the proportion of petroleum used and by a substantial decrease in the contribution of coal. In the short run immediately after the first oil shock, however, the fall in the petroleum-GNP ratio resulted not only in an increase in the electricity-GNP ratio but also in an increase in the coal share in total energy consumption (see Figure 2.9). The overall energy savings and, in particular, petroleum savings, together with the introduction of various kinds of labor-saving automation, brought about an increase in electricity consumption.

This phenomenon has extremely important implications in analyzing the future relationship between the GNP and electricity consumption. Moreover, it should be taken into consideration that the future increases in real petroleum prices for the next 10 to 15 years will be less than 3 to 4 percent of the projected increase in real wage rates in industrialized countries, according to general consensus on the forecasts.

The eighth factor in Table 2.8, on the other hand, shows changes in industrial structure between 1960 and 1979 for the three countries. It is noticeable

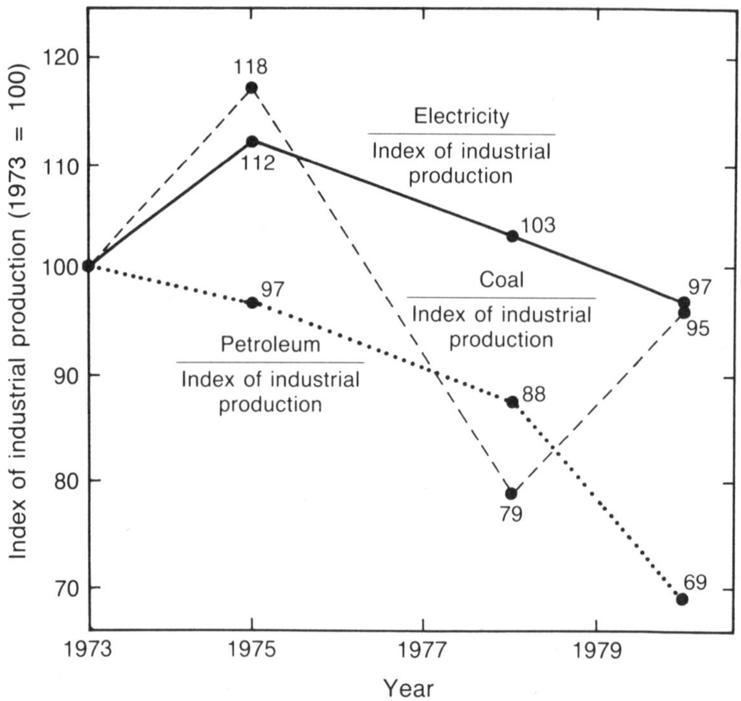

Figure 2.9. Historical changes in coal, petroleum, and electricity consumption per unit of industrial output for Japan. Note the substantial reduction of petroleum use during the 1970s.

that the output share of energy-intensive manufactures in Japan increased from 33 percent in 1965 to 37 percent in 1970, remained at the same level in 1975, and slightly declined to 36 percent in 1980. In Korea the proportion of electricity-intensive manufactures in total industrial growth output sharply increased from 14 percent in 1960 to 36 percent in 1970, remained at the same level in 1975, and again rose to 37 percent in 1980.

Table 2.12 shows the detailed breakdown of the structural changes that occurred in the manufacturing sector of Japan between 1965 and 1979. These data reveal that Japan has begun to transform its industrial structure from the electricity-intensive industry stage into the technology-intensive industry stage. From these international comparative analyses of the structural changes and electricity-based technological advances, it can be postulated that industrial sectors classified under energy-intensive industry coincide with those of

Table 2.12
Composition and Growth of the Manufacturing Sector in Japan, 1965–1979

Subsector	Composition (%) 1965	1970	1975	1979	Annual Growth Rate (%) 1965–1979
Light manufactures					
Food, beverages, tobacco	18.6	12.8	13.2	12.1	4.4
Textiles and apparel	9.8	7.4	6.6	5.6	3.4
Lumber and wood products	4.8	4.1	3.8	3.2	4.7
Miscellaneous manufactures	7.7	7.0	6.8	6.9	6.9
Subtotal	(40.9)	(32.3)	(30.4)	(27.8)	(4.8)
Power-intensive manufactures					
Pulp and paper	3.8	3.6	3.3	3.4	6.8
Chemicals, coal, petroleum	12.2	13.4	14.5	14.2	8.8
Non-metallic mineral products	3.5	3.8	3.4	3.5	7.6
Primary iron and steel	11.5	13.7	13.4	12.3	8.2
Primary non-ferrous metals	2.1	2.3	2.2	2.3	8.4
Subtotal	(33.1)	(36.8)	(36.8)	(35.7)	(8.3)
Technology-intensive manufactures					
Fabricated metal products	3.6	4.2	4.3	4.4	9.2
General machinery	7.3	9.9	9.5	10.1	10.1
Electric machinery and equipment	5.0	7.3	7.4	10.2	13.4
Transport machinery	9.1	9.4	10.5	10.3	8.6
Precision and optical products	1.0	1.1	1.1	1.5	10.9
Subtotal	(26.0)	(31.9)	(32.8)	(36.5)	(10.3)

Sources: Ke et al. (1983); Kim et al. (1983); Kimura et al. (1983).

the electricity-intensive industry, and that there may exist a transitional stage of industrialization between the energy-intensive industry stage and the technology-intensive industry stage, namely, the electricity-intensive industry stage. These observations also suggest that Korea is in the process of advancing to the electricity-intensive industry stage.

In summary, the phenomenal growth in the number of households served by electricity supplies, combined with the rapid increase in real income and the decline in relative price of electricity, turned out to be the main determinants of the increase in residential electricity consumption between 1960 and 1980 in China (Taipei), Japan, and Korea. At the same time, the growth in industrial production, changes in industrial structure, and technological advances were the most important determinants of the increase in industrial demand for electricity. In addition, government policy toward future industrial development and social changes will significantly influence future electricity consumption. On the other hand, the price variable was insignificant at the macro level, but a much more detailed study of particular industries would be necessary to establish the precise role of the price variable.

Finally, in considering the time lag of adjustment by household and business consumption behavior to electricity price increases and the time lags of the structural adjustments by the entire economy to lower economic growth, it is desirable to analyze in detail the movements of various energy data and related economic indicators for several years after the second oil shock, especially the 1981–1984 period. This is necessary to evaluate more accurately both the electricity price elasticity at the stage where many adjustments are near completion, and the partial income elasticity foreseen in the long run after eliminating the demand-suppressing effect of electricity price.

FORECASTING METHODS AND RESULTS

A variety of forecasting methods are available and used by Japanese utilities, Korea Electric Power Corp. (KEPCO), and Taipower. They include judgment, trend extrapolation, time series, econometric, end-use, and combinations of these methods. Each forecasting method has certain strengths and weaknesses as well as resource requirements, such as data, time, cost, and technical capabilities required for implementation. These characteristics of major methods are summarized in Table 2.13. In general, expert judgment must be used in all forecasting. For the other methods, expanded capability is generally accompanied by increased resource requirements for conducting the analysis.

Experiences in the past decade clearly demonstrate that conventional forecasting methods are inadequate to deal with sudden changes in energy systems. Considerable advances have since been made, however, in electricity demand forecasting methods including progress in time series (Bolet, 1985), in econometric analysis (Jorgenson, 1982; Sagawa and Kibune, 1983), and

Table 2.13
Comparison of Major Forecasting Methods

			Resource Requirements			
Method Description	Strength	Weakness	Data	Time	Cost	Technical Capabilities
(1) Judgmental						
Expert opinion	Low cost Particularly useful for intangibles	Qualitative Difficult to replicate Difficult to find experts	Low	Low	Low	Knowledge of power system; interview skills
(2) Analytical						
Trend analysis	Simple Low cost	Neglecting underlying influences	Low	Low	Low	Statistical analysis; computer
Time series analysis	Extracting more information from data	Neglecting underlying influences	Medium	Medium	Medium	Advanced statistical analysis; computer
Econometric analysis	Explicitly analyzing underlying causes of trends and patterns	Can be costly due to data requirement Requires experience to use and interpret results	Medium/ high	Medium/ high	Medium/ high	Econometrics; computer
End-use analysis (engineering process)	Easy to derive intuitive understanding of analytical results	Very expensive and data intensive	High	High	High	Energy-economic analysis; computer

in energy-economic models such as PILOT (Dantzig, 1982, 1985) and ETA-MACRO (Manne, 1977, 1983). In addition, a number of highly sophisticated end-use models have been developed by the Electric Power Research Institute in the United States (Cambridge Systematics, 1982) and by Tokyo Electric Power Company in Japan (Sato, 1984).

In Japan long-term forecasts are prepared by the Japan Electric Power Survey Committee (JEPSC), which uses both econometric and end-use models. An end-use model is primarily utilized for midterm projections (usually five to nine years), and its results are checked by an econometric model. For longer-run projections, an econometric model is primarily used, whereas an end-use model is used for checking purposes. In general the econometric model utilizes statistical methods and historical data to produce a forecast. The consumption of electricity is linked to such factors as economic, demographic, and conservation conditions. As a consequence, the model captures the historic behavior pattern of electricity consumers. Because it relies on historical data, the econometric model may not be appropriate to forecast the effects of new technologies, of radical departures from past patterns of consumption behavior, or of structural changes in the economy.

The end-use approach (engineering process model), on the other hand, utilizes detailed information on the end use of electricity. As a result, the engineering model allows for very detailed descriptions of the customer response to new or speculative technologies, changes in price structures, and income. A major problem with the engineering model is the availability of detailed empirical data. These two approaches differ greatly, but it is apparent that the use of the two models practiced in JEPSC could exploit the complementary strength of each, thus improving the forecast made by the individual utilities.

Unlike Japanese utilities, both KEPCO and Taipower basically utilize simple econometric models to forecast the energy requirements for lighting and the power requirements for commercial and industrial customers. In these models, the consumption of electricity is linked to the GNP, sectoral value added, and time. A comparison of current forecasting methods used by Japanese utilities, KEPCO, and Taipower is summarized in Table 2.14.

Historic evolution of forecasting. It is both interesting and instructive to compare the historic evolution of forecasting methods in these utilities. A summary of the key methods used and relevant system characteristics for the years 1950, 1960, 1970, and 1980 are shown in Table 2.15. This table indicates that the sophistication and complexity of forecasting methods parallel the growth of the power system, the complexity of the economy, and the perceived uncertainty of future changes.

In the 1950s both KEPCO and Taipower were relatively small in size, their systems were expanding rapidly, and the overwhelming majority of their load was in heavy industries. Thus judgment and trend analysis were adequate methods for forecasting. Even the larger system in Japan was just beginning to use the time series method.

In the 1960s the power systems in all three countries expanded with steady economic growth. Time series analysis became the dominant method for forecasting. In the late 1960s, as economic prosperity continued in these countries, residential and commercial demand for electricity began to grow rapidly. Simultaneously, industrial loads diversified from concentrated heavy industries to more diffuse light manufacturing. Consequently, econometric analysis started to assume importance.

Until the 1970s the principal determinants of electricity demand growth in all these countries, including demographic change, economic growth, economic structure, energy prices, energy technologies, and energy use patterns, have been relatively stable or predictable. The 1973 oil embargo and the subsequent energy crunches, however, have altered the situation dramatically. Almost all the important determinants have large uncertainties either in themselves or in the degree of their impact on electricity consumption. Table 2.16 summarizes the changes that are the root causes for the growing difficulty in electricity demand forecasting. These changes have had strong effects on the development of forecasting methods. First, macroeconomic forecasting in the past was separate and distinct from electricity demand forecasting. With these changes, the distinction blurred, and more and more electricity demand forecasting was then made within the overall context of macroeconomic forecasting. Second, these changes have had large impacts on end use in the form of energy conservation and economic structural change. As a result, there is a growing emphasis on end-use models. The Japanese utilities have already developed comprehensive end-use models, whereas Taipower is actively studying economic structural change.

These changes and uncertainties in the determinants of electricity demand have affected not only the development of forecasting methods but perception of future demand growth. For example, whereas in the past the growth of the GNP and electricity have been closely coupled in both time and magnitude, in the future the relationship between these factors is expected to be modified to a considerable degree by conservation, new technology, structural changes in the economy, and changes in consumption patterns. Specifically, in 1980 for the first time Japanese utilities experienced a negative growth rate of electricity consumption, although real GNP grew at a positive rate. The identical phenomenon occurred in China (Taipei) in 1981. Unlike its two neighbors, in 1980 the demand for electricity in Korea increased at 5 percent, whereas the GNP in real terms recorded a negative growth for the first time in Korea's history. This instability in the relationship between the GNP and electricity, although perhaps temporary, suggests that factors other than simple GNP and time trend and more complex interactions between factors will need

Table 2.14
Characteristics of Demand Forecasting Models Practiced in Japan's Utilities, KEPCO, and Taipower

	Japan's Utilities[a]	KEPCO	Taipower
Primary type	End-use model with economic submodels	Econometric model	Econometric model
Model variable	Economic indicators Business plans in various industries Conservation Appliances technologies Time series data	GNP Time series data	GNP Time series data
Range of period	5–9 years	Long-term	Long-term
Data source	Market survey by a market research company	GNP projections by EPB[b]	GDP projections by CEPD[c]
Categories	Lighting Commercial Small power (< 500 kW) Large power (> 500 kW) Other	Residential Industrial Commercial	Lighting Eight different industries, e.g., construction, commerce, manufacturing
Use of results	Check with econometric models Reconduct market survey if necessary Revise annually	Check with the trend of more advanced cross-sectional models Revise annually	Construct three scenarios from different assumptions Pick out one scenario by consulting the market survey of large power customer (> 1,000 kW) Revise annually

Supplementary procedures	Internal econometric models by JEPSC[d] Electric power sector model by CRIEPI	Market survey on large power customer (> 500 kW)	Market survey on residential small power, large power customers Consult the status of electrification in more advanced countries
Problems	Future uncertainties Model the industrial structural change Model new technology innovation, e.g., office automation, air conditioner	Future uncertainties Model the industrial structural change Model new technology innovation, e.g., office automation, air conditioner Integrate econometric end-use model Credibility of market survey data	Future uncertainties Model the industrial structural change Model new technology innovation, e.g., office automation, air conditioner Integrate econometric end-use model Credibility of market survey data

a. The methodologies described here are quite close to, but not exactly equivalent to, the official ones in Japan's utilities.

b. Economic Planning Board.

c. Council of Economic Planning and Development.

d. Japan's Electric Power Survey Council.

to be considered to be forecasted in the future. As Table 2.17 shows, these changes in demand determinants have already significantly affected the perception of future growth in the utilities of all three countries.

Table 2.18 presents long-term projections of socioeconomic indications, total energy, and electricity in China (Taipei), Japan, and Korea. In these projections, China (Taipei) assumes a 5.9 percent annual average growth rate of real GNP during the period 1980–2000, compared with 9.6 percent during the period 1960–1980. Japan assumes 4.5 percent compared with 7.6 percent, and Korea assumes 6.9 percent versus 8.6 percent. Future projections of GNP growth, compared with the past 20 years, have been substantially reduced in all three countries. Accordingly, in China (Taipei) the annual growth rate of power consumption is projected to decline sharply from 13 percent in the period 1960–1980 to 5 percent in the period 1980–2000; in Japan from 8.6 to 3.9 percent; and in Korea from 19.1 to 9.2 percent. As a result, the GNP elastic-

Table 2.15
Comparison of Forecasting Methods and System Characteristics

Country/Item	1950	1960	1970	1980
Korea				
System load (GW)	0.1[a]	0.4[a]	1.4	5.5
Demand growth rate (%/year)	20.0[a]	20.0[a]	15.0	1.9
Key forecasting methods	Trend analysis	Trend analysis/ time series	Time series/ econometrics	Econometrics
China (Taipei)				
System load (GW)	0.2[a]	0.7	2.1	6.7
Demand growth rate (%/year)	20.0[a]	14.0	15.3	10.4
Key forecasting methods	Trend analysis	Trend analysis/ time series	Time series/ econometrics	Econometrics
Japan				
System load (GW)	4.0[a]	17.0	49.0	89.0
Demand growth rate (%/year)	18.0[a]	15.5	15.5	-2.5
Key forecasting methods	Trend analysis/ time series	Time series	Econometrics	Econometric/ end-use

a. Estimate.

ity of electricity consumption will fall from 1.4 to 0.85 in China (Taipei), from 1.14 to 0.87 in Japan, and from 2.2 to 1.3 in Korea. Finally, in the year 2000 per capita electricity consumption is expected to range from about 3,800 kWh in Korea to 4,500 kWh in China (Taipei), and 8,700 kWh in Japan.

Table 2.16
Major Changes in Demand Factors

Demand Factor	Condition Pre-1970s	Condition Post-1970s
Electric rates	Continual decrease in real terms due to scale economies and supply technology improvements	Sign of real increases and future uncertainties due to probable exhaustion of scale economies and large escalations in capital and fuel costs
Fuel prices	Generally stable oil prices	Large step increases and uncertainties in oil prices that have not only produced shock effects to the economy but also increased public awareness to energy conservation
Economic growth	Generally upward trend since 1945	Stagflation and uncertainty
Economic stability	Generally stable growth trend since 1945	Greater uncertainty
Economic production	Expansion of service sector in most industrialized countries	Continuation of trend with uncertain impact due to information technology developments
Demographic	Varying among different countries	
Public attitudes	Increasing awareness of environmental effects of economic growth	Added awareness of energy conservation
Government policies	Widespread support for electricity growth	Overall emphasis on energy conservation
Technical changes	Continual technological advances	Uncertain prospect of information technology development and its impact on electricity demand

How accurate have they been? In general, during the 1970s and early 1980s, the annual forecasts published by the three utility systems have been consistently lower each year. This is illustrated by Figure 2.10 for all nine Japanese utilities, Figure 2.11 for KEPCO, and Figure 2.12 for Taipower. Note that not only are the forecasted curves displaced to the right as demand fails to come up to previous forecasts, but the slope of the forecasts has a downward trend as well. This, as discussed in more detail in Chapters 3 and 4, has led to some difficulties in capacity planning and load management.

It is valuable to explore why past forecasts have been generally high. It should be first pointed out, of course, that there is an inherent bias in utility projections toward the high side. This is a natural outgrowth of the utility aversion to capacity shortages. Overestimates are seen to be preferable to underestimates because they leave the utility the option of delaying planned capacity expansions. Shortages are physically and politically more difficult to handle.

Table 2.17
Changes in Long-Term Forecasts

Country and Year of Forecast	Load Forecast (GW) for Year 1990
Korea	
1968	25
1970	25
1980	20
1981	19
1982	16
1983	16
China (Taipei)	
1973	26
1975	12
1978	20
1980	17
1982	10
1984	12
Japan	
1981	176
1982	157
1983	132
1984	128

Sources: Ke et al. (1983); Kim et al. (1983); Kimura et al. (1983).

The Japanese utilities have done an analysis of the factors that led to overestimates in the 1980s. As shown in Figure 2.13, one predicted growth rate between 1979 and 1982 was 5.6 percent, although the actual rate was slightly on the negative side of zero. The difference was attributed to increased technical efficiency (0.9 percent), increased public awareness (1.1 percent), structural changes in industry (2.2 percent), economic stagnation (1.6 percent), and the remainder from minor deviations in weather patterns. Projecting these trends out to 1992, as indicated in Figure 2.13, shows that the difference in projections amounts to about 40,000 MW. In other words, without the considerations of the changes listed above, instead of the approximately 140 GW forecasted in 1983 for 1992, a total of nearly 180 GW peak capacity would have been needed. This is a huge difference, representing more than $60 billion of investments over ten years, and illustrating the difficulty and importance of forecasting.

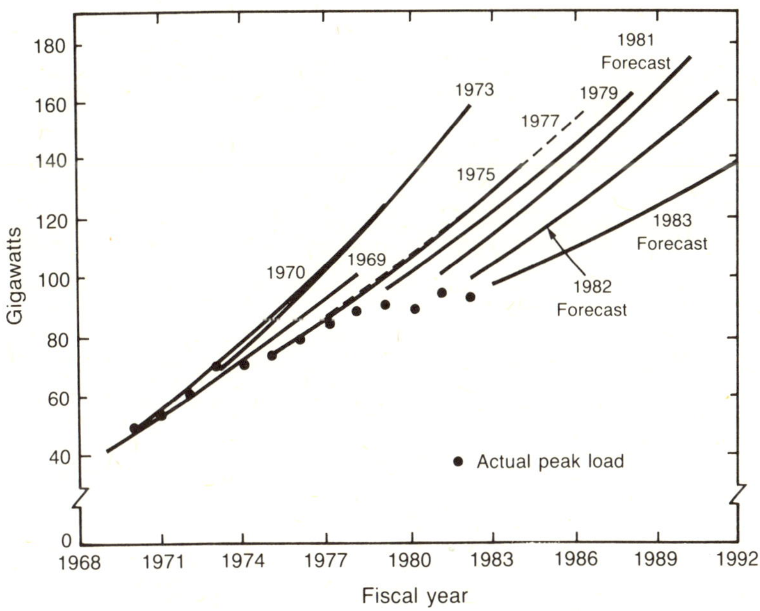

Figure 2.10. History of peak load forecasting in Japan. Note that succeeding forecasts have nearly always been lower than earlier ones.

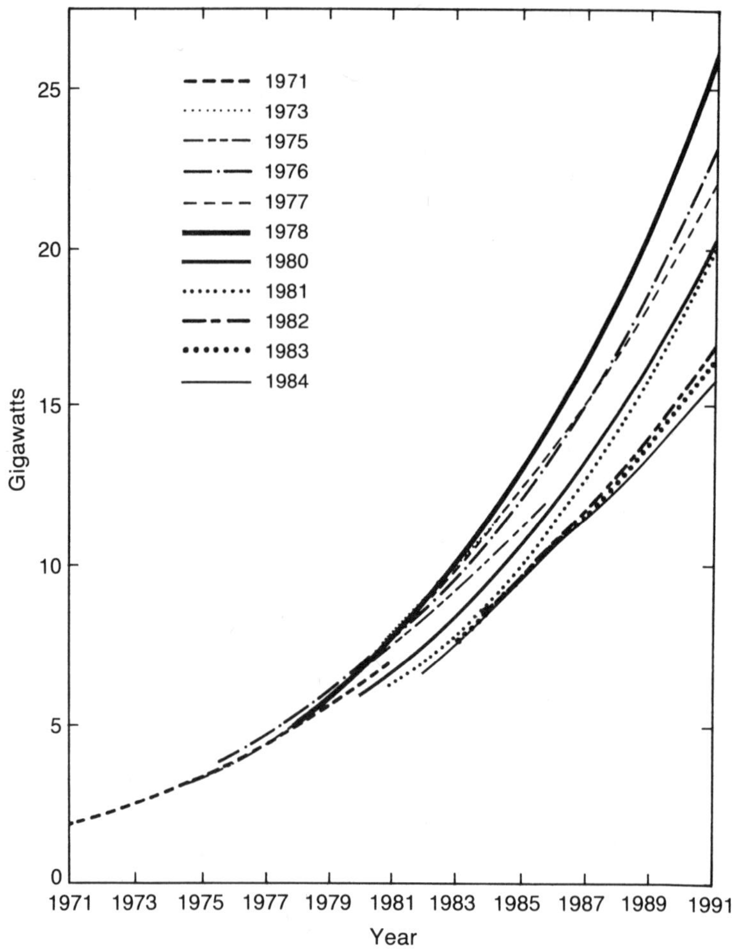

Figure 2.11. History of peak load forecasting in Korea. Note that later forecasts are substantially lower than earlier ones.

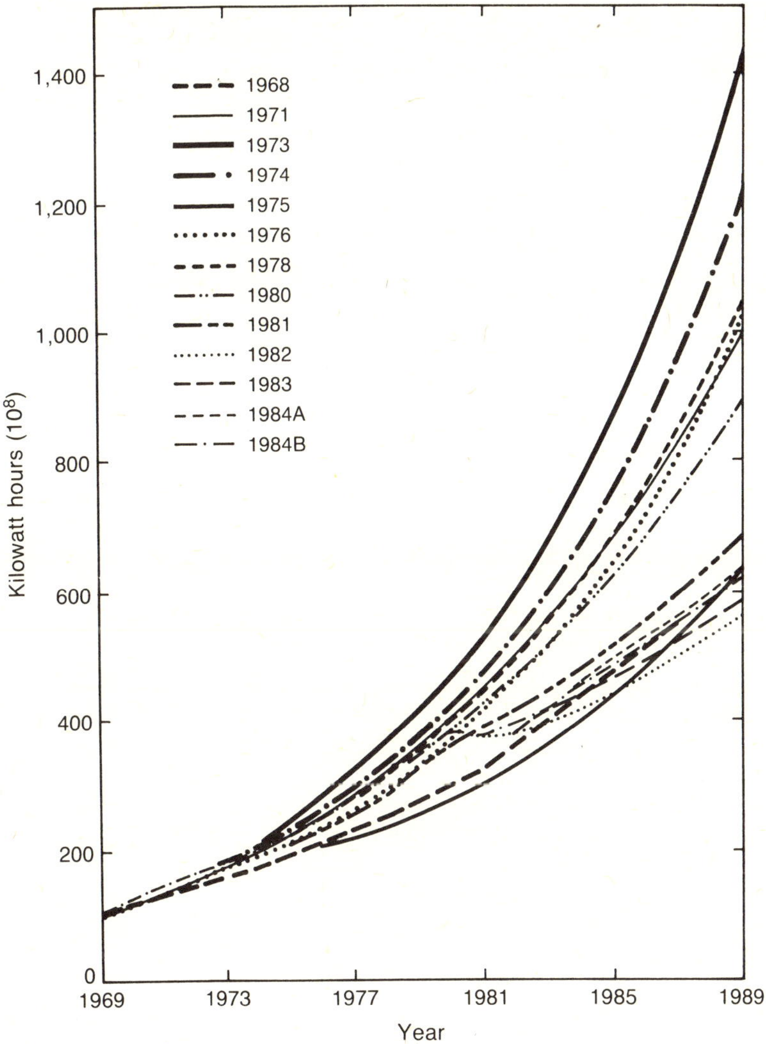

Figure 2.12. History of electricity demand forecasts in China (Taipei). Note the substantial lowering of forecasted future demand that occurred through the 1970s.

POLICY IMPLICATIONS

Contemplation of these historical experiences and changing prospects in electricity demand forecasting results in the following policy implications for the development and application of forecasting methodology:

1. *Electricity demand forecasting should be coordinated and integrated with international economic analysis.* This need is particularly acute for the Northeast Asian countries because of their heavy dependence on foreign natural resources and international trade. As the availability of natural resources becomes increasingly politicized and the competition in world trade becomes increasingly fierce, such coordination and integration of electric demand forecasting with international economic analysis are imperative in order to narrow the uncertainty faced by electric utilities.

2. *Electricity demand forecasting should be further coordinated with national economic policy.* In all three countries, government economic policies ex-

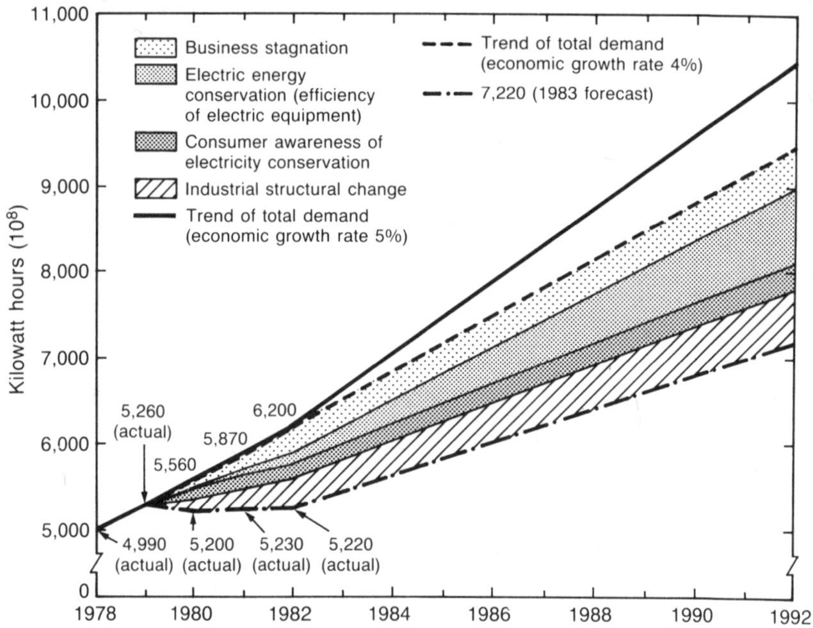

Figure 2.13. Forecasted and actual Japanese electricity demand. The shaded areas are labeled with the various reasons why actual demand was less than forecasted. Conservation and increased efficiency had the largest impact.

ert strong influences on future economic development, energy prices, and consumption patterns. In the past, there has been considerable coordination between electric utility forecasting and national macroeconomic analysis. As international trade becomes increasingly competitive, electricity consumption becomes an even more important part of national economic planning. Thus, further coordination and integration are recommended.

3. *Electricity demand forecasting should integrate econometric analysis with end-use analysis.* As the economy of each country changes with shifts in production pattern and resource availability, end-use analysis becomes increasingly important. There has been a growing effort in all three countries to develop detailed end-use models. Such models, however, have the tendency to be myopic in predicting future technological innovation and lack of consistency and interaction among different end uses. Therefore, econometric analysis needs to be integrated with end-use models to provide a check-and-balance in demand forecasting.
4. *Electricity demand forecasting should include uncertainty or risk analysis of alternative futures.* Although the purpose of forecasting is to reduce future uncertainty or risk, it is impossible to understand all interactions among the relevant determinants of electricity demand; forecasts are thus inherently uncertain and risky. It is important to include an analysis of the residual uncertainty and its impact on utility decisions and risk in demand forecasting.
5. *The resources used for electricity demand forecasting should represent a balance between its costs and the benefits derived.* As discussed in the beginning of this section, forecasting requires resources in terms of data, time, money, technical skills, and facilities. Therefore, the development and application of forecasting methods should always strive to balance costs and benefits in terms of the uncertainty reduced and decisions affected.

In summary, between 1960 and 1982, there have been important changes in the structure of electricity demand in Northeast Asia. These include erosions of the historically strong correlations between electricity consumption and economic growth, slowdowns in industrial load growth, and peak load growth outstripping baseload growth. Thus, in addition to income and energy prices, in forecasting the major determinants of electricity demand should include economic structure, technology development, and public policy. In addition, it should attempt to systematically present risk and uncertainty and seek to achieve a balance between the cost and benefits of forecasting itself. Because of these changes, traditional demand forecasting techniques such as trend extrapolation and time series analysis are no longer adequate.

Table 2.18
Electric Power Forecasts, 1960–2000

	China (Taipei)	Japan	Korea
Projections of socioeconomic indicators			
Growth rate of real GNP (%)			
1960–80 (actual)	9.6	7.6	8.6
1980–2000	6.0	4.5	6.9
2000–2020	—	—	5.0
Growth rate of real per capita GNP (%)			
1960–80 (actual)	7.1	6.4	6.7
1980–2000	4.6	4.0	5.7
2000–2020	—	—	3.6
Proportion of real GNP (%)			
Industry			
1980	61	46	62
2000	61	48	64
2020	—	—	66
Transportation			
1980	6	6	6
2000	7	4	7
2020	—	—	7
Commerce and services			
1980	33	48	32
2000	32	48	29
2020	—	—	27
Population growth (%)			
1980–1990	1.5	0.5	1.5
1990–2000	1.0	0.4	1.3
2000–2020	—	0.0	1.2
Size of households (persons/household)			
1980	4.8	3.2	4.8
2000	3.9	2.8	3.9
2020	—	—	3.5
Urbanization rate (%)			
1980	70	76	67
2000	88	—	81
2020	—	—	85

Table 2.18 (*continued*)

	China (Taipei)	Japan	Korea
Energy projection			
Total energy demand (103 TOE)			
1980	25,700	375,000	43,900
1990	39,800	554,000	86,900
	(4.5)[a]	(3.2)	(7.1)
2000	65,400	489,000	145,000
	(5.0)	(2.7)	(5.3)
2020	—	—	268,000
			(3.1)
Energy-GNP elasticities			
1960–1980	1.07	1.01	0.95
1980–1990	0.74	0.64	0.93
1990–2000	0.85	0.68	0.79
2000–2020	—	—	0.62
Sectoral composition (%)			
Industry			
1980	72	62	43
1990	69	59	66
2000	69	55	65
Transportation			
1980	11	16	13
1990	12	15	13
2000	12	15	12
Residential and commercial			
1980	17	22	21
1990	19	26	23
2000	19	31	23
Energy projection by type (%)			
Coal			
1980	15	17	—
1990	29	20	—
2000	29	19	—
Petroleum			
1980	70	66	—
1990	45	49	—
2000	44	38	—

Table 2.18 (*continued*)

	China (Taipei)	Japan	Korea
Gas			
1980	6	6	—
1990	3	12	—
2000	7	11	—
Hydro			
1980	2	6	—
1990	4	5	—
2000	3	5	—
Nuclear			
1980	7	5	—
1990	18	11	—
2000	14	18	—
New energy			
1980	0	0.3	—
1990	1	4	—
2000	3	10	—
Power projection			
Total power demand (GWh)			
1980	39,200	520,000	32,700
	(13.2)	(8.6)	(19.1)
1990	66,100	795,000	82,200
	(5.4)	(4.3)	(9.6)
2000	115,000	1,120,000	190,000
	(5.7)	(3.5)	(8.7)
Per capita consumption (kWh/person)			
1960	321	1,060	62
1980	2,220	4,450	850
	(10.2)	(7.4)	(14.8)
2000	4,990	8,740	3,790
	(4.1)	(3.4)	(7.7)
Power-GNP elasticities			
1960–1980	1.38	1.13	2.22
1980–1990	0.88	0.87	1.33
Sectoral composition (%)			
Industry			
1980	68	64[b]	70
1990	68	61[b]	65
2000	66	58[b]	62

Table 2.18 (*continued*)

	China (Taipei)	Japan	Korea
Transportation			
1980	1	b	1
1990	2	b	2
2000	2	b	2
Residential			
1980	18	63[c]	16
1990	22	39[c]	20
2000	24	42[c]	23
Commercial			
1980	13	c	13
1990	8	c	13
2000	8	c	13

Sources: Ke et al. (1983); Kim et al. (1983); Kimura et al. (1983).

a. Figures in parentheses are percentage growth rates over previous years.

b. Japanese industry and transportation figures are combined.

c. Japanese residential and commercial figures are combined.

— Not available.

NOTES

1. In analyzing the changes in demand characteristics of the residential and commercial sectors, it is necessary to analyze in detail the demand data after the second oil shock, taking into consideration the effects of both the time-lag of the response by residential-commercial demand to the increase in energy prices since the first oil shock and the energy-saving technological progress of home electric appliances.

2. It should be remarked that if percentage increases in residential electricity consumption *per capita* instead of *per household* are compared, the resulting income elasticity of residential consumption would become much larger, because between 1960 and 1980 the growth rate in the number of households served by electricity supplies exceeded considerably that in population. See items 2 and 3 of Table 2.8.

REFERENCES

Baxter, R. E., and Rees, R. (1968). "Analysis of the Industrial Demand for Electricity." *Economic Journal* 78:277–298.

Bolet, A. M., ed. (1985). *Forecasting U.S. Electricity Demand: Trends and Methodologies.* Boulder, CO: Westview Press.

Cambridge Systematics, Inc. (1982). *Residential End-Use Energy Planning System (REEPS)*. EA-2565. Palo Alto, CA: Electric Power Research Institute.

Dantzig, George (1982). *PILOT-1980 Energy-Economic Model.* EA-2090. Palo Alto, CA: Electric Power Research Institute.

——— (1985). *PILOT-1983 Model.* Technical Report SOL 85-12, Department of Operations Research, Stanford University, Stanford, CA (July).

Fisher, F. M., and Kaysen, C. (1975). *The Demand for Electricity in the United States.* Amsterdam: North Holland.

Hawkins, R. G. (1975). "The Demand for Electricity: A Cross-Section Study of New South Wales and the Australian Capital Territory." *Economic Record* 51:1–18.

Houthakker, H. S., Verleger, P. K., and Sheehan, D. P. (1974). "Dynamic Demand Analyses for Gasoline and Residential Electricity." *American Journal of Agricultural Economics* 56:413–418.

Jorgenson, Dale (1982). *Energy Supply Restrictions: Energy Systems and Economic Effects.* EA-2553. Palo Alto, CA: Electric Power Research Institute.

Manne, Alan (1977). *ETA-Macro: A Model of Energy-Economic Interactions.* EA-592. Palo Alto, CA: Electric Power Research Institute.

——— (1983). *ETA-Macro: A Progress Report.* EA-3170. Palo Alto, CA: Electric Power Research Institute.

Sagawa, Naoto, and Kibune, Hisao (1983). "International Comparison of Energy Demand Structures." *Energy in Japan.* Tokyo: The Institute of Energy Economics.

Sato, Norikazu (1984). "Japanese Electric Power Demand." Presented at the Second Working Group Meeting on the Electric Future of the ASEAN Countries, Korean Electric Power Company, Seoul.

Taylor, L. D. (1975). "The Demand for Electricity: A Survey." *Bell Journal of Economics* 6:74–110.

CHAPTER 3

LOAD MANAGEMENT

R. Tabors and M. Castillo Bonet

The electric power system is unique in the energy supply sector from several perspectives. First, it operates in real time: the product—electricity—must, under most circumstances, be consumed at the same time that it is produced. Second, power systems are capital-cost-intensive structures that are subject to some economies of scale. Finally, and not uniquely, electric power systems are made up of both the supplier and the consumer. Uniquely, however, because of the real time nature of the system, the actions of supplier and consumer are intimately bound together. The implications of this reality, unfortunately, are too often forgotten by utility planners.

This neglect is now being remedied in many utilities (see, e.g., Pacific Gas & Electric, 1983). A representative statement of the need for this change is that of Paul Zimmer, chairman of the Edison Electric Institute (EEI):

> Very exciting, too, is a resurgence of attention to the needs and desires of our customers. For a number of years, we have been preoccupied with the problems of building new plants, finding affordable fuels, and meeting overall load growth. In this process, wc tended to lump customers together and think of them solely in terms of the demand they represented. Now, more and more companies are readdressing their customers as individuals with differing energy needs, and we are increasingly presenting customers with choices in electricity use. To my mind, this is an exciting return to basics, and a healthy direction for the industry. (*Electrical World,* 5/84, p. 30).

This chapter first examines the interaction between electricity suppliers and customers in Northeast Asia. It follows the history of the match between supply and demand in the three countries and the load management practices that have been used to try to bring them into synchronization.

The second section looks at new methods in load management as a means of introducing additional demand-side control into the utility operating systems. It reviews current methods and technology in load management and looks to areas where additional attention may be focused in the future. Tariffs are discussed in detail in Chapter 6, and so only their specific uses as a tool in load management are discussed here.

The final section summarizes the critical issues in load management for Northeast Asia in the decade ahead. It looks at the patterns of power system development that could result from the region's rapidly growing utilities having substantially different load management options and information available to them than established systems had during their periods of rapid growth.

THE MATCH BETWEEN SUPPLY AND DEMAND IN NORTHEAST ASIA

Before examining the history and prospects for load management in Northeast Asia, it is useful to present a brief review of the principles and terms used in load management analyses of utility operations. In doing so, we use examples from the utilities under study in this book. (Readers already familiar with load management principles and terms may wish to skip to Load and Its Management in Northeast Asia on page 71.)

Utility Operations

An electric power system must be capable of meeting the load (demand) given the available generating capacity and other system control options such as load management. The most generally accepted criterion is that the load should be met in such a way as to produce electricity at the lowest cost possible at acceptable levels of reliability.

For the system operator, the cost of electricity is defined only in terms of operating cost. The system operator inherits a stock of capital and a set of load management options and must work within these constraints. In addition, the system operator works to maintain a prespecified level of reliability.

Reliability is an important parameter throughout the energy sector, particularly in electric power systems. In power systems, where there are several measures of reliablity, the most common measure of generation *system* reliability is the *loss of load probability* (LOLP). This is the statistical expectation of the number of days per year (or month) when an outage due to generating capacity shortage could occur. It does not directly indicate the reliability at the point of consumption, however. In nearly all countries, the unreliability of the transmission and distribution (T&D) systems, from storms for example, is much greater than that due to insufficient generating capacity (a factor of 10 to 20 times greater in the United States, for example).

The reliability criterion is set on the basis of economics, safety, and convenience (Munasinghe, 1979). All else being equal, a high LOLP is costly because of loss of production and the backup power that customers would need to provide for themselves. A very low LOLP, on the other hand, is expensive because of the extra generating capacity needed to assure that enough will always be available. Thus there is an optimum LOLP where the total costs of the utility and the customer are minimized. The LOLP criterion can range from three days per year to one day in ten years in different countries. The LOLP goal for the Korea Electric Power Corporation (KEPCO) is 0.7 days/year and for Taipower is 3 days/year. In Japan, however, the Tokyo Electric Power Corporation (TEPCO) states its LOLP is 0.3 days/month and believes that setting monthly LOLP targets provides for more consistent year-round service.

Historically, generation reliability has been increased by increasing the margin of excess capacity in order to be able to cover any conceivable combination of demand during the breakdown or maintenance of generating plants. There are two common measures of the degree of extra capacity—*planning* and *operating reserve margins*. The first is simply the capacity in excess of peak load divided by the peak load. The operating reserve margin takes into consideration that not all facilities may be available to meet peak load, owing to such contraints as seasonal water flow or maintenance. A typical planning reserve margin is 20 percent on a yearly basis with operating reserve somewhat lower. Recently in Korea, however, because demand growth was lower than anticipated, operating reserve has increased to more than 20 percent. Reserve margins can also be expressed on a monthly basis. TEPCO's LOLP, for example, is met if the operating reserve margin is 8 percent on a monthly basis.

Load management techniques are based on the premise that true reliablity—at the level of the consumer—is more closely and cheaply attainable by consideration of changes across the entire system, from generation, through transmission and distribution, to consumption (Morgan and Talukdar, 1979).

The world of a systems operator can be defined in three time blocks. The first, from milliseconds to seconds, is the time frame in which frequency control within a power system takes place automatically. The second time frame of the system operator is that of minutes, generally 10 to 60. This is the time frame of system dispatch. During this period the operator adds units to the system in order to meet expected load, and changes the relative loading points of individual plants both for serving the load and for guaranteeing system reliability through spinning reserves.

Spinning reserves fall within this second time frame. These reserves are generally defined as the amount of capacity that can be brought on line within 15 minutes. The size of the spinning reserve is a function of the size of the largest unit in operation among other factors. For example, when Korean nuclear units (KNU) 1-3 were the largest plants in its system, KEPCO maintained a 650 MW spinning reserve. Now that it has a 950 MW nuclear unit (KNU-5), however, its spinning reserve has also been increased.

The third time frame in which the operator works is that of hours to days, the time frame of unit commitment. Here the operator is concerned with units available for operation—that is, which units are not down for planned maintenance. The operator is concerned that scheduling over the next hour to the next week offers sufficient availability to guarantee reliability—that is, the operator must be sure that no large plants will be down during the time period for maintenance that cannot be covered by other facilities on line.

A utility typically has swings in demand that must be met by the load-following capabilities of the system. These shifts occur on a daily, weekly, seasonal, and annual basis. Figure 3.1 illustrates two of these shifts in load patterns for Taipower. It also illustrates the change in the shape of the daily load for a typical weekday in summer over a period of 20 years and shows that there has been both a general increase in the total demand as well as a change in the shape over the day. An example of a change in the daily pattern is the emergence of the lunchtime "valley" in demand over the period.

Whereas Figure 3.1 combines changes both in absolute and relative demand, Figure 3.2 examines the changes in load shape alone. This figure, again from Taipower, shows the evolution of the peak summer day over 17 years, but in reference to the daily average. It also shows that the relative size of the minimum (off-peak) demand has decreased over the period.

For many purposes, it is useful and important to know the composition of power plants being used to meet demand because some power plants are much more expensive to run than others. Figure 3.3, an example from KEPCO, shows the composition of power plants being used on average for the whole year and for typical days in summer and winter. (Note the different vertical scales.) The dispatch order for the system can be seen from the bottom to the top and begins with an even base of nuclear power, then thermal high-efficiency systems, thermal low-efficiency systems, hydro, internal combustion engines in the summer season, and finally pumped hydro. The loading order as described is normally based on the marginal operating cost of the individual units. In this case, therefore, nuclear has the lowest marginal cost and the internal combustion engine the highest.

Utilities divide power plants into three general categories according to their physical and economic characteristics. Peaking units have low capital costs, short start-up times and generally high fuel costs, and operate less than 20 percent of the year. Baseload plants have low fuel costs, frequently have high capital costs, work best when operating continuously, and generally operate 73–100 percent of the year. Intermediate plants have characteristics that lie between those for peaking and baseload plants.

The relative costs of these three types of plants according to the percentage of the year they operate are shown in Figure 3.4A. Peaking units, for example, are able to produce cheaper electricity than the other categories only if they operate less than t_r hours per year. Figure 3.4B shows what is often referred to in discussions of utility performance as a "load duration curve." It plots the power being generated as a percentage of the peak annual demand against the percentage of the time that demand is required. If the demand were constant all year, the curve would be a straight line, indicating demand to be 100 percent. Only baseload plants would then be needed. It also shows how the load patterns of baseload, intermediate, and peaking plants differ.

Figure 3.5 shows the results of these calculations for KEPCO's system in 1983. Costs are calculated on the basis of long-run marginal costs, that is,

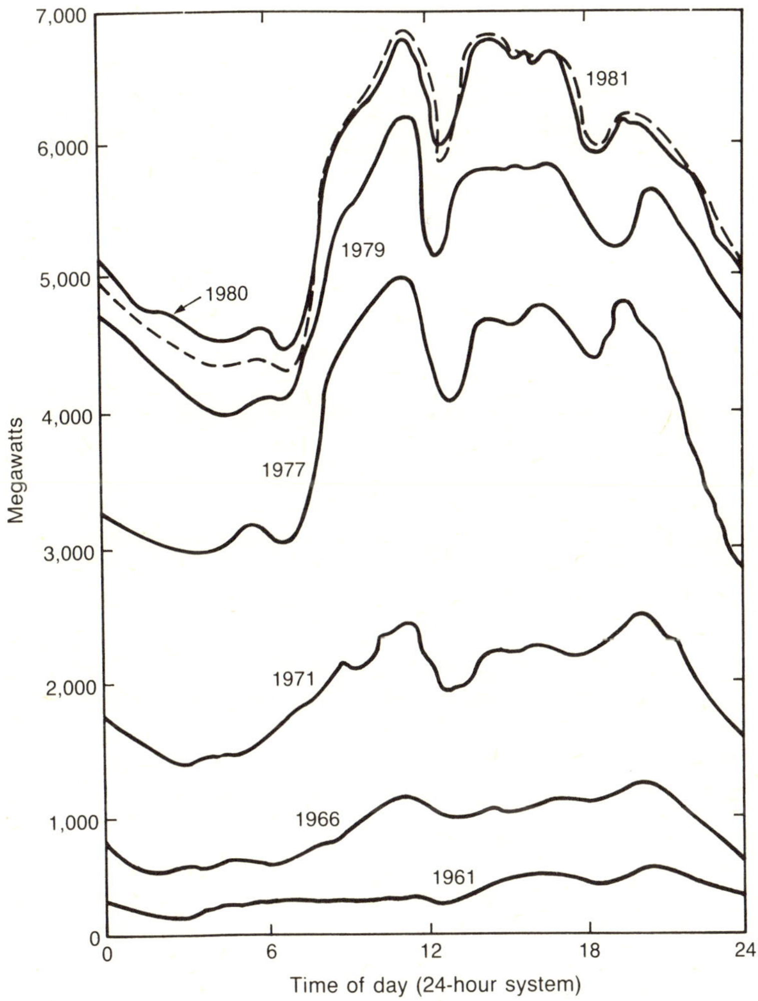

Figure 3.1. Historical trends of summer day load patterns for Taipower. Note the increasingly prominent afternoon peak over time.

the additional cost of supplying an increment of electricity over the long term. It thus includes capital as well as operating costs in each category of demand. (Precise definitions of this term and others used in this chapter can be found in Munasinghe and Warford, 1982.)

A popular measure of power system economics is the *load factor,* defined as the fraction that area under the load duration curve represents of the total area. All else being equal, utilities prefer to have higher load factors because this allows them to operate their capital-intensive facilities to a greater extent, thus lowering average costs.

Load factors are quite different among different classes of customers as shown in Figure 3.6. In general, residential customers have the lowest annual load factors, followed by commercial customers who also have the largest variation within classes, and the industrial customers, with the highest annual load factors. It is clear, therefore, that the composition of customers will affect load factors as will climate, which also affects the need for heating and cooling. Since both the customer mix and the climate tend to be different

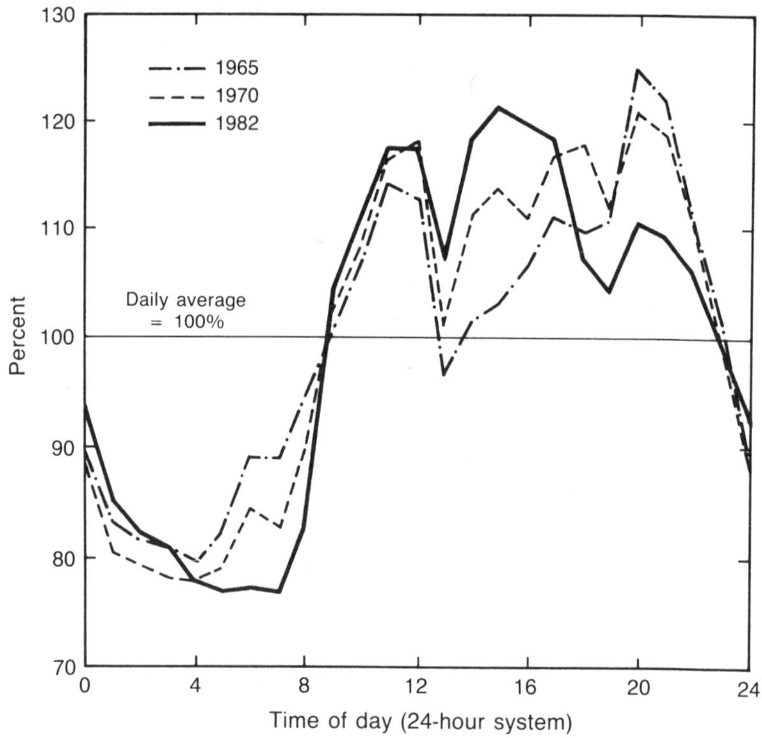

Figure 3.2. Historical comparison of peak summer day load patterns for Taipower.

for countries at different levels of development, direct comparisons of load factors should be done with care.[1]

Figure 3.7 shows a pair of load curves from Taipower. It can be seen that at least 70 percent of full demand was required 50 percent of the time in 1973. The fact that the curve for 1982 was generally lower than the one for 1975 indicates that there was a decrease in the load factor—that is, the system became less efficient.[2]

Figure 3.7 also illustrates that the important operational characteristics of utility can change over a relatively short period of time. In this case, conditions have deteriorated from the perspective of the system operator. The inset in Figure 3.7 shows the "needle peak," which marks the capacity required to meet the most demanding 1 percent of the yearly load. The needle peak in Figure 3.7 has changed in a negative direction from the perspective of systems operations. The energy content of the needle peak (measured as the area under the curve) has dropped to 0.03 percent, whereas the capacity requirement to serve that needle peak has increased to 4.50 percent of maximum capacity.

As illustrated by these examples, the challenge of systems operation is to deal with a commodity that is both produced and consumed in essentially the same instant. This means that the load and the supply system must be matched on a minute-by-minute basis. It is thus apparent that load management should be considered an additional tool to balance supply and demand.

LOAD AND ITS MANAGEMENT IN NORTHEAST ASIA

Load management[3] refers generally to an effort on the part of the utility to influence the pattern of demand in such a way as to lower overall costs or to increase overall reliability for the system. As such, the term *load management* covers a multitude of actions that range from physical control over customer demand units to economic signals that offer incentives for customers to modify their demand (up or down). The techniques and equipment for implementing load management have been evolving rapidly. These trends and their implications for Northeast Asia are discussed in the next section.

In recent years the utilities of Northeast Asia have found their load factors and other indicators of the match between demand and supply to be stagnant or indicating a decline in efficiency relative to their system design. As a result, and in concert with utilities in much of the world, load management techniques have started to become more prominent in their planning and operations. In this section, the present load management practices of the three countries are described in the context of their history and of the match between supply and demand.

Figure 3.8 shows the recent history and projected future of TEPCO's load factors. It illustrates a declining trend both historically and in forecasts. This

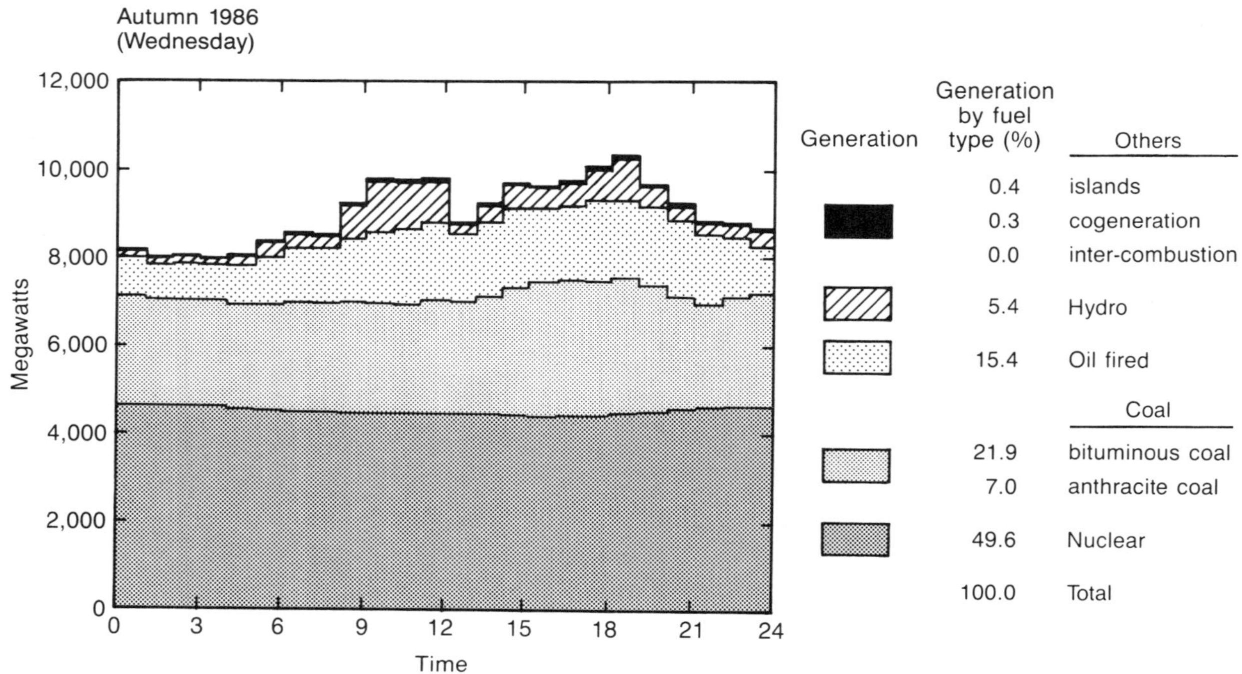

Figure 3.3. Typical daily load patterns in Autumn 1986 (above) and Winter, Spring and Summer 1987 (below) for KEPCO. Also shown is the amount of generation by each fuel type.

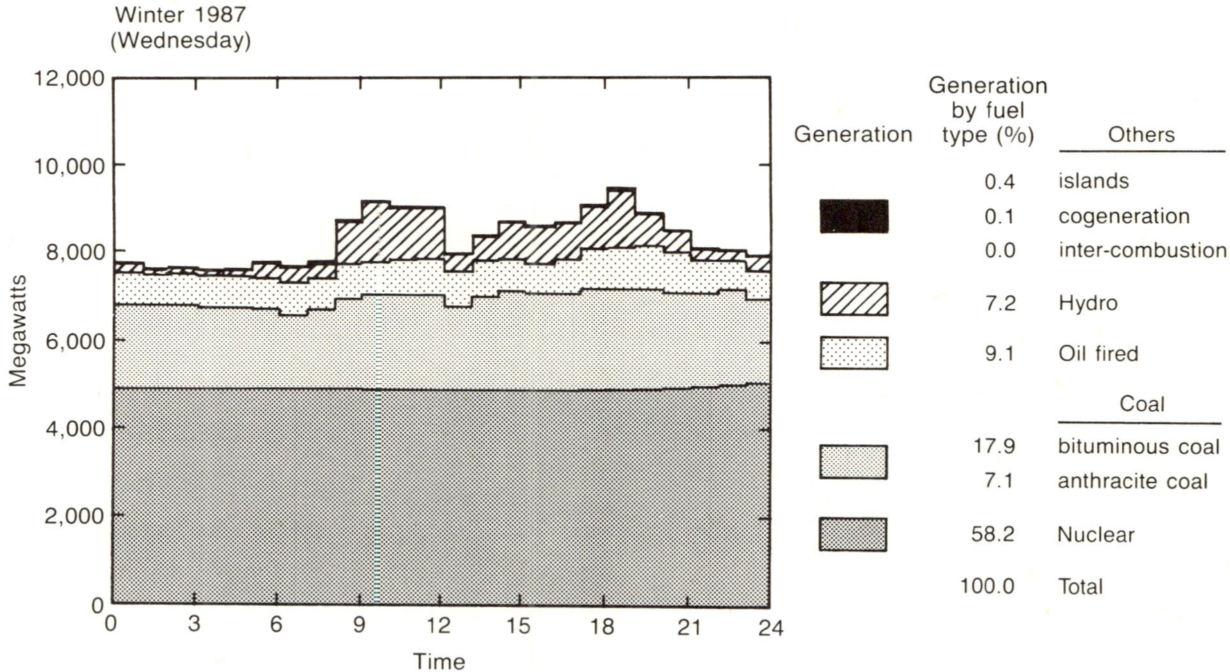

Figure 3.3. *(continued)*

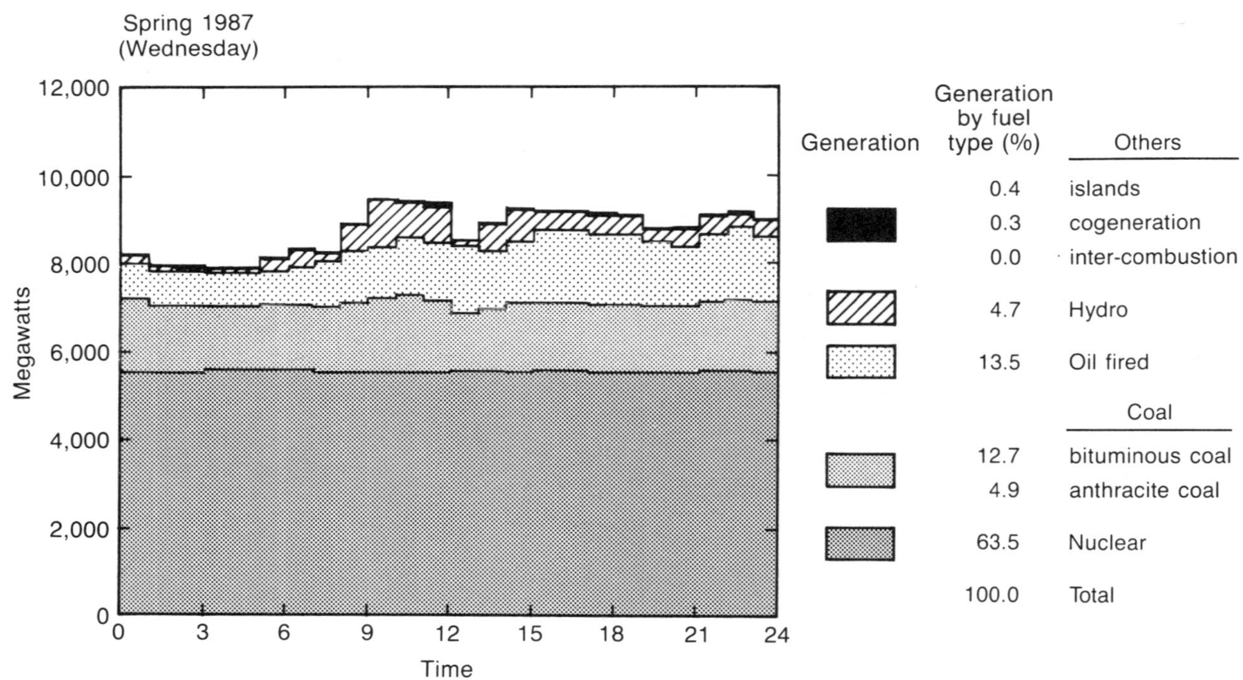

Figure 3.3. *(continued)*

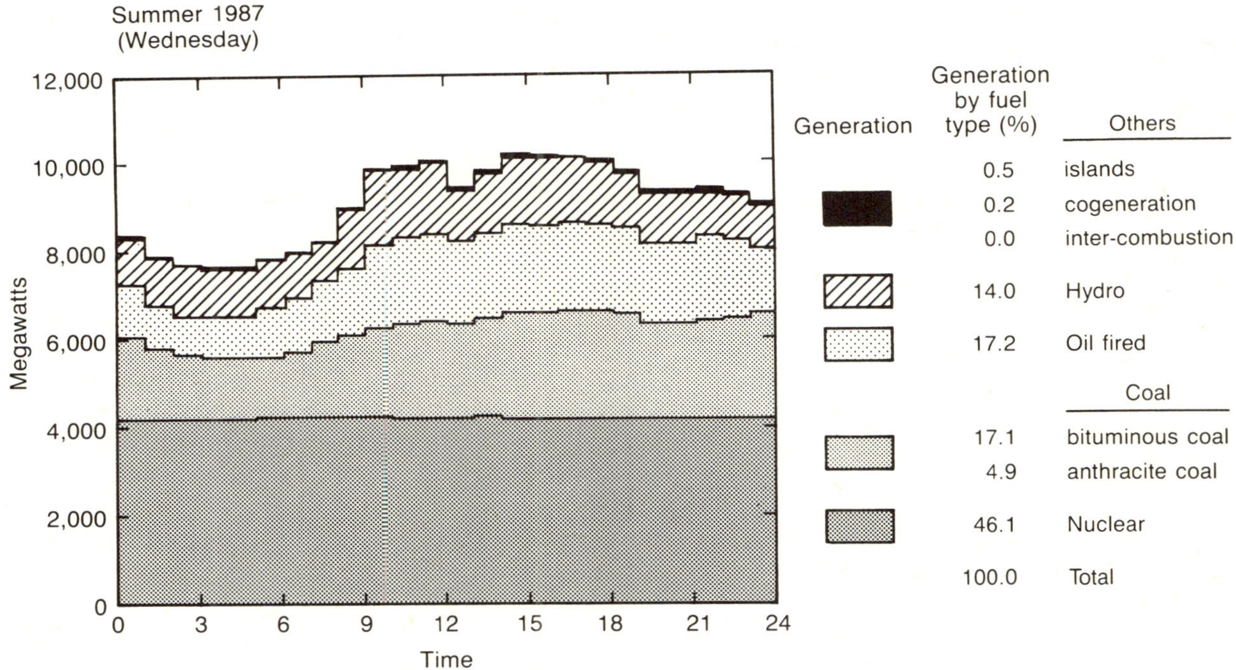

Figure 3.3. *(continued)*

change (for the worse in Northeast Asia) may in part be due to an increasing block tariff structure[4] that did not have systematic time-differentiated tariffs during a period of highly variable demand. This has in part led TEPCO to its "special rates" (see Tariff Structures in Chapter 6), an initial step in load management techniques. TEPCO also relies on pumped storage as a means of adapting supply to demand. Japan is a pioneer in pumped storage technology, having introduced it in 1931, well before most other countries in the

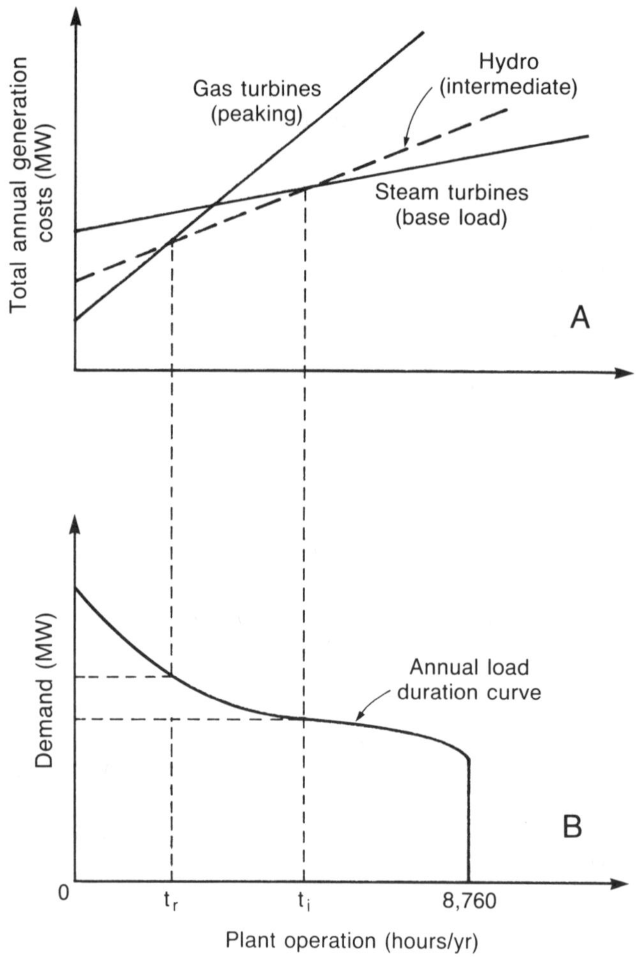

Figure 3.4. Theoretical relationships between load durations and costs for baseload, intermediate, and peaking plants (see Munasinghe and Warford, 1982; also see Figure 3.5).

region. The cost of new sites for pumped storage are shifting the relative economics of this strategy to favor load management over storage alternatives.

Within TEPCO, some load management is induced by the use of interruptible-service contracts and time-of-use rates. Nevertheless, these tariffs are applicable primarily to large industrial consumers as "special tariffs." They fulfill a number of functions, only one of which is load control. Special tariffs result from proprietary negotiations between the utility and individual customers (approved by the Ministry of International Trade and Industry, or MITI) to which access is limited, making outside assessment difficult. The types of special rates that were in effect at TEPCO and the type of response anticipated are summarized below.

The interruptible rate schedules are similar to those in the United States, the United Kingdom, and France. Rates are set according to the customer's willingness to remove all or a portion of the load given a fixed advance time.

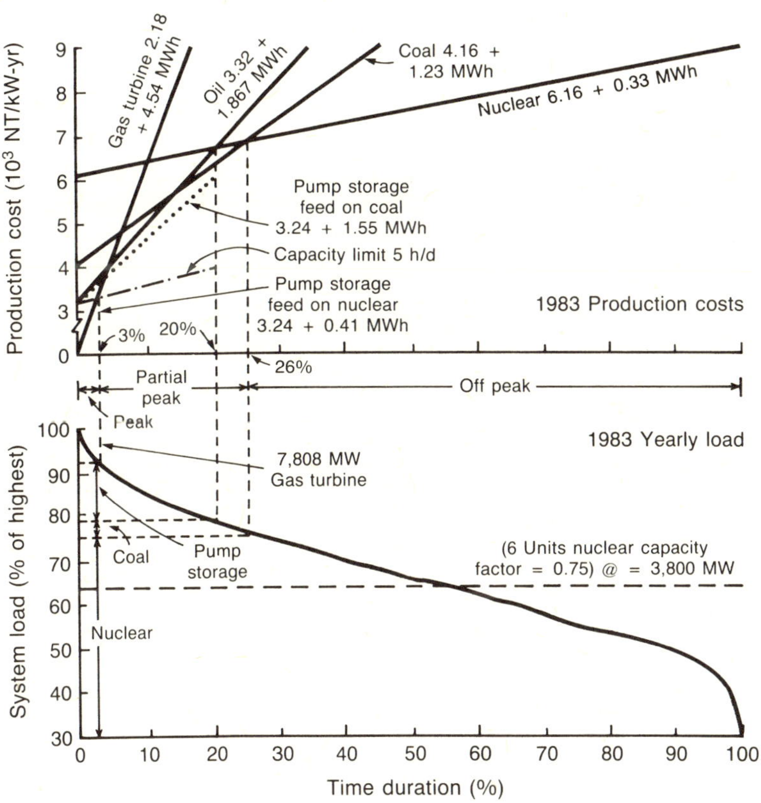

Figure 3.5. Actual relationships between loads and costs for KEPCO in 1983. (Also see Figure 3.4.)

As in the United States and Europe, these rates are used for emergency conditions and were accepted by industries with high usage but relatively insensitive loads. Electric arc furnaces and mining operations for minerals such as phosphate, which use electrically powered drag lines, are examples of industries that can respond best to interruptible rates.

Load modification rates are more difficult to characterize, but they are based on a customer's ability to shift load. They tend to be fixed in advance to offer a bonus to customers who will change their behavior. Thus they are based on a customer's pattern of consumption under the rate relative to the customer's pattern of consumption before the rate was introduced. Actual implementation of "what-would-have-been" rates is difficult. In Japan the actual amount of benefit received by a customer and by the utility is not known by other actors in the system; it is known only by the utility and customer affected. These rate structures seem to be well accepted and they appear to work effectively. Customers did modify their loads at times of system peaks and did move demand to valleys through these contractual equivalents to the more common time-of-use rate structures found in Europe and to a lesser extent in the United States (Tabors, 1985; Meta Systems, 1987).

The Japanese system has been peaking in the summer since the late 1960s as shown in Figure 3.9. This is primarily due to the increased penetration of air-conditioning in the residential and commercial sectors, as discussed in Chapter 2. Table 3.1 shows the growth in peak and baseload demand during the 1970s. Its seasonal component seems to have a large growth potential. In fiscal year 1973, for example, market penetration of air conditioners

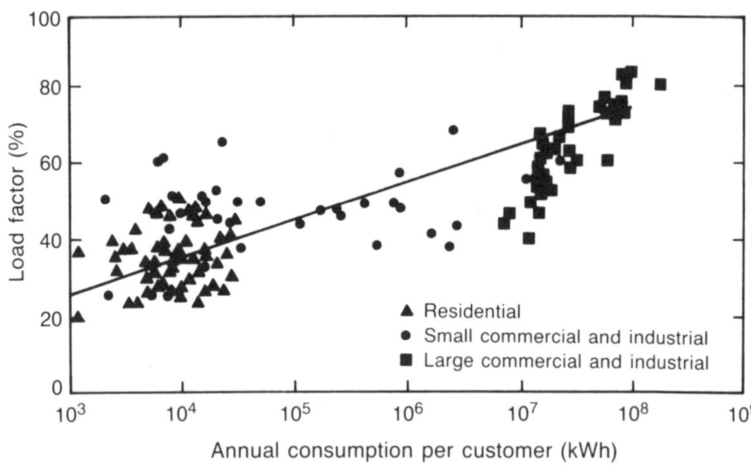

Figure 3.6. Annual load factor for customers of different types. Note that residential and smaller commercial customers typically have lower load factors (see Schneider, 1982).

in Japan was only 18 percent, whereas by 1981 rapid diffusion was evident in Tokyo but was still only 60 percent.

Figure 3.10 shows that, in contrast to Japan, Korea's load factor has started to decline only since 1980. Before 1981 the winter peak in Korea coincided with the annual peak, but that year they were about equal and since then the summer peak has dominated (Figure 3.11). As a result, in late 1977 KEPCO decided to introduce a time-of-use rate structure applicable to large

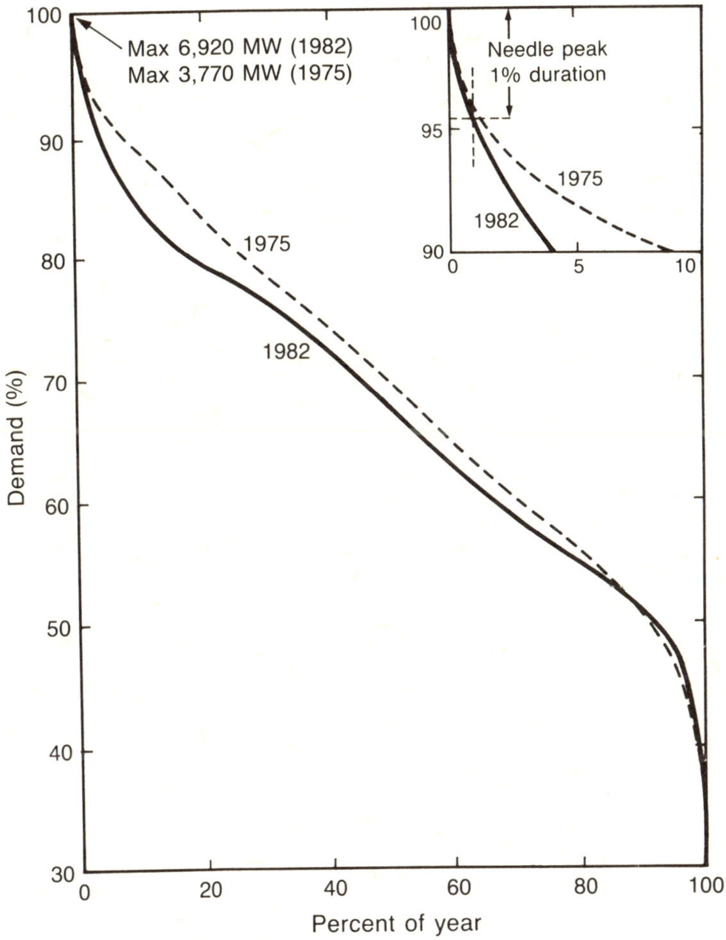

Figure 3.7. Annual load duration curves for Taipower in 1975 and 1982. Note increased narrowness of needle peak.

industrial users (above 300 kW of contracted maximum demand). That rate structure was modified somewhat during the following years (see Table 3.2), but in essence it remains the same.

In the future, the summer peak is expected to continue to dominate in Korea. The factors that lead to this conclusion are:

1. Rapid expansion of residential and commercial cooling load (the annual growth rate in air-conditioning load has been nearly 20 percent in the 1980s).
2. Decreased impact of peak pricing due to longer duration of cooling load.
3. Decrease in nighttime and weekend loads owing to relative wage increases and decreases in working hours per week.
4. Television broadcasting (perhaps) in the afternoon.

Taipower's load factors also began to decrease when its system began summer peaking in the early 1970s (Figure 3.12), and by 1980 there was a marked difference between the two seasonal peaks as shown in Figure 3.13. This summer peak was due to the rapid increase in the use of air conditioners that accompanied the large growth of income per capita in China (Taipei), which, as Figure 3.12 shows, is expected to continue in the future. The use of air-conditioning in the residential and public service sectors was late in starting, but in the late 1970s it began to catch up with the commercial sector. In addition, the relatively low cost of electricity and lack of appliance standards (low average efficiency) led to rapid diffusion of air conditioners. This is expected to continue because saturation in all three sectors was only about 20 percent in the early 1980s.

It is interesting to note that although the income and electricity consumption of China (Taipei) and Korea lag behind that of Japan by 15 and 20 years,

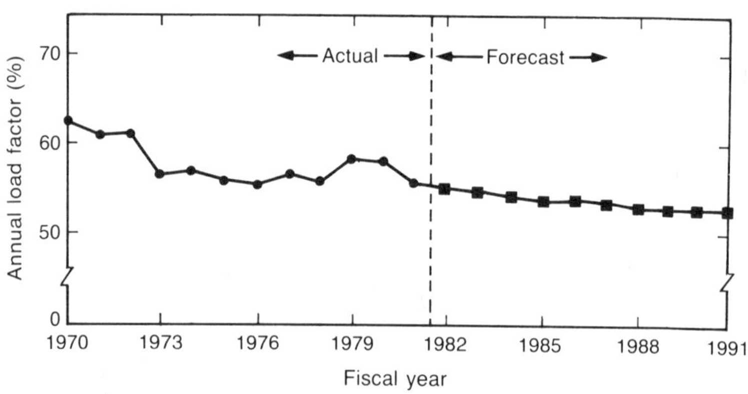

Figure 3.8. Historical and forecast load factors for TEPCO in 1982.

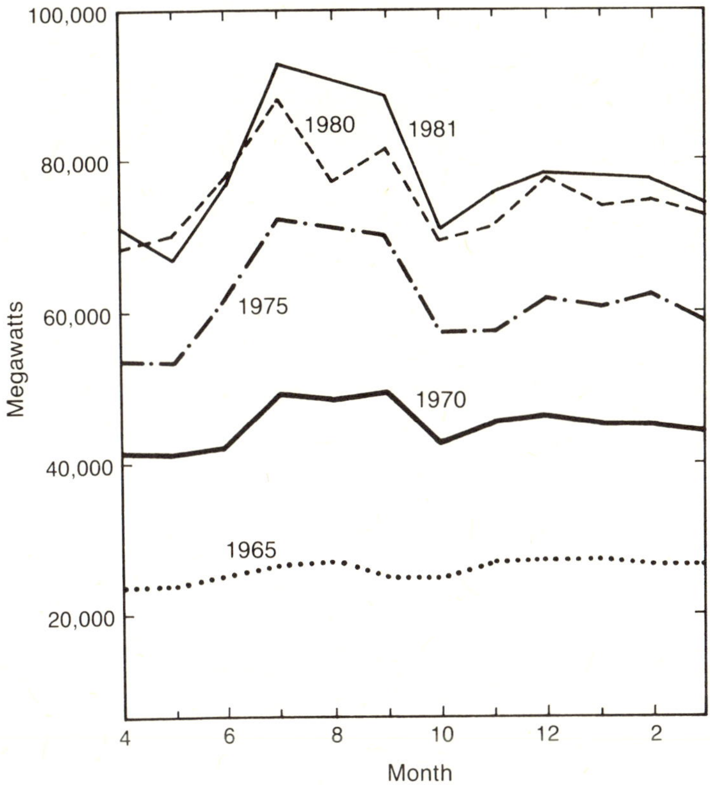

Figure 3.9. Historical trend of monthly maximum demand of all nine Japanese electric utilities.

respectively, their cooling demands lag by only 4 and 13 years. This is due to the climatic differences and, presumably, the demonstration effect.

The period of worst impact of air-conditioning load is Taipower's needle peak, which occurs in the hottest period of the year (see Figure 3.7). This needle peak is exacerbated by the declining block tariff structure, thus encouraging additional consumption at lower rates. In the early 1980s, as Figure 3.7 and Table 3.3 show, Taipower experienced prominent annual needle peaks. It is instructive to calculate the net cost of meeting such a peak. Table 3.4 demonstrates how the gross cost of generating the first 1 percent of hours is derived. The social cost of curtailment in power-intensive industries is roughly equal to their product values of NT$16/kWh. Net cost becomes NT$48/kWh, which is expensive electricity. Clearly some application of load management would be worthwhile. (See What the Future May Bring later in this chapter.)

Another factor negatively affecting future load factors in Japan is its changing mix of industries. The trend is away from heavy industries, such as iron, steel, and aluminum, that tend to operate 24 hours per day seven days per week. More of the economy is being directed to industries such as electronics and machinery fabrication, which operate mostly during business hours, thus decreasing the relative baseload and increasing peak.

In Korea the late evening peak is due in part to the contribution of residential and commercial lighting. The time-of-use rate is, however, applied only to the big industrial users who are better able to adapt to this type of rate structure. Allocation of generating cost to more nearly reflect contribution to peak is accomplished by relatively high (flat) rates for commercial users. Residential customers appear to be insufficiently charged in recovering these costs.

The evolution of Taipower's summer daily load curve for the past 20 years is shown in Figure 3.1. The load factor is becoming increasingly low due to the increase of residential and commercial loads and to the concentration of the industrial load during the standard working hours. The latter is partly due to the increase of noneconomic benefits to labor (fewer nighttime and weekend shifts). This situation has been exacerbated by the reindustrialization policy of China (Taipei), which is to switch away from industries that tend to have a baseload schedule and toward high-tech light manufacturing. This is taking place at a time when industrial demand in general is depressed by the world economic situation and residential demand in China (Taipei) continues to grow at a fast pace.

Taipower introduced a time-of-use rate structure in 1980 for industrial users above 500 kW of contracted demand (users with a contracted capacity between 100 kW and 500 kW have the option to choose time-of-use rates), which was modified in 1981 to increase the gap between off-peak and peak prices. This effort has had some impact on load profiles as shown in Table 3.5. Additional information on load shape is, however, required. The response to the price signals could be confused with the effect of reduced demand from the baseload industries.

The effect of the time-of-use rates on the demand schedule of Korean industrial users can be observed in Figure 3.14 for the case of some major industrial sectors. The drop in demand during the peak hours is very apparent. On the other hand, lighting is one of the chief components of demand in residential apartment buildings because coal and propane are the fuels primarily used for cooking in the urban centers of Korea. Lighting demand, of course, is not appropriately managed by time-of-day rates.

Another load management issue is the large drop in demand during the lunch-break period (noon to 1 p.m.) that occurs in the KEPCO, TEPCO, and Taipower systems on weekdays. In Korea this drop is associated exclusively with the demand of the industrial sector, where job shifts have a relatively rigid schedule, and it does not seem to be related to the commercial sector.

Table 3.1
System Peak Load in Japan, 1975–1981 (MW)

Item	1975	1976	1977	1978	1979	1980	1981
Summer load	6,990	7,560	7,979	9,300	8,250	7,900	9,910
Basic load	15,020	16,520	16,740	17,600	19,000	19,380	19,650
Total	22,010	24,080	24,710	26,900	27,250	27,280	29,560

Notes: Summer seasonal air-conditioning load contribution; maximum three-day average peak load at transmission end.

Figure 3.1 shows this dip in Taipower's demand, which amounts to almost 800 MW and creates significant problems for the system operators.

Unlike its neighbors, China (Taipei) has experienced relatively recent periods of frequent power shortages caused by power plant failures. Under emergency government orders in the years prior to 1979, energy-intensive industries such as aluminum, carbide, caustic soda, steel, textiles, and cement (in that order of priority) were required to curtail usage, upon Taipower's request, by 10–20 percent. Their demand charge was refunded on a straight-line scale basis. About 100 customers using over 500 kW were curtailed each time, reliev-

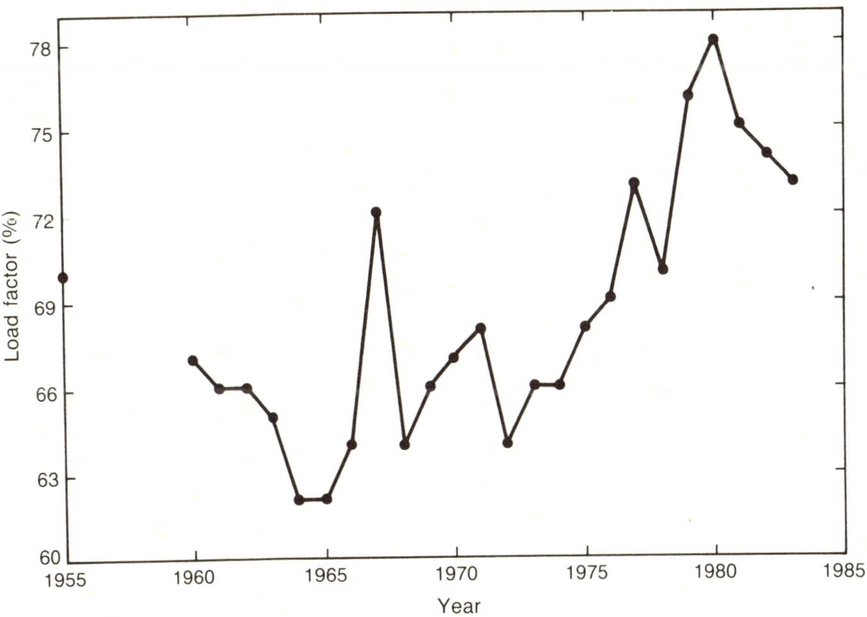

Figure 3.10. History of annual load factors for KEPCO.

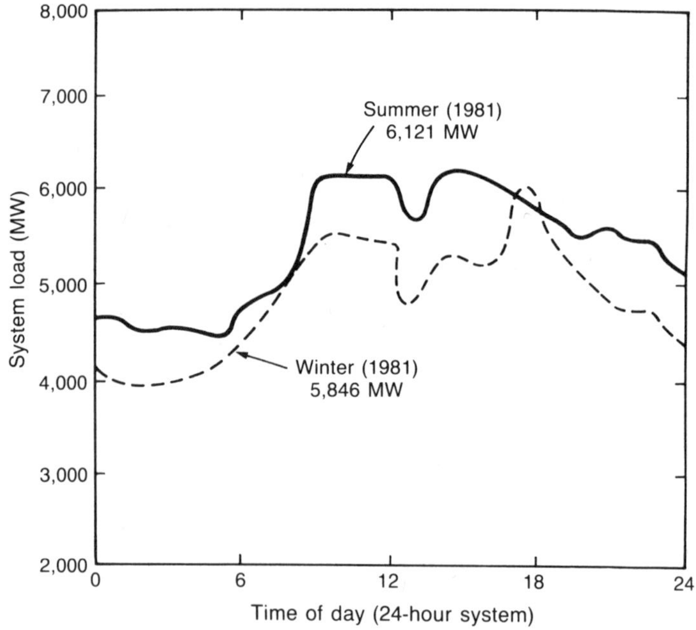

Figure 3.11. Peak winter and summer daily load patterns for KEPCO in 1981.

Table 3.2
a. Relative Price Ratios for the Time-of-Use Rate Structure in Korea

Date Effective	Off-Peak	Intermediate	Peak
1 December 1977	1.0	2.0	5.0
12 June 1979	1.0	2.5	7.0
1 February 1980	1.0	2.0	5.0
1 December 1981	1.0	1.5	3.0

b. Time-of-Use Rate Schedule in Korea

Time Period	Summer (March through September)	Winter (October through February)
Off-peak	10 p.m. to 6 a.m.	10 p.m. to 6 a.m.
Partial peak	6 a.m. to 7 p.m.	6 a.m. to 6 p.m.
On-peak	7 p.m. to 10 p.m.	6 p.m. to 10 p.m.

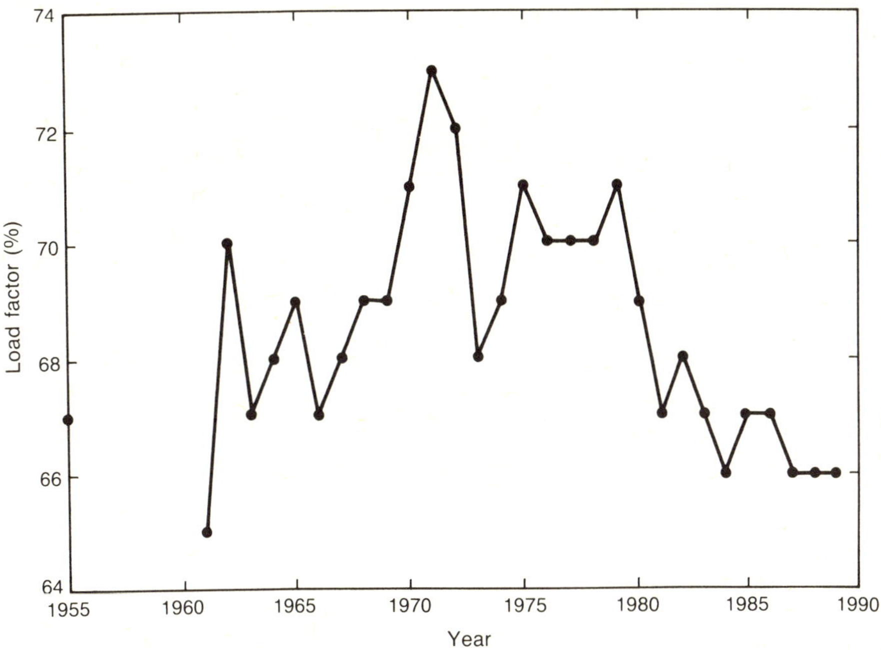

Figure 3.12. Historical and forecasted load factors for Taipower in 1982.

ing 100 MW of peak load. As Figure 3.15 indicates, the tight years in the late 1970s contrast sharply with the excess capacity margins of the 1980s.

To cope with serious power shortages during the summer of 1979, Taipower executed rotational shutdowns, under government authorization, for two workdays each month in industries with more than a 500 kW load. On prearranged days, these industries had their demand meters read twice by Taipower—early in the morning and late in the evening. Thus each customer was evaluated with respect to shutdown effectiveness and was refunded double the demand charge on a straight-line basis. With some 1,200 customers affected, about 300 MW were curtailed during weekdays. This was equivalent to 5 percent of the system peak load.

LOAD MANAGEMENT SYSTEMS: CURRENT TRENDS

The modern concept of load management in the electric power industry has developed only since the mid-1970s. It has been only recently, therefore, that some utilities have begun to see their roles as extending beyond just the supply of electricity to customers at whichever times and in whatever amounts the customers might demand. Previously, the matching of demand and sup-

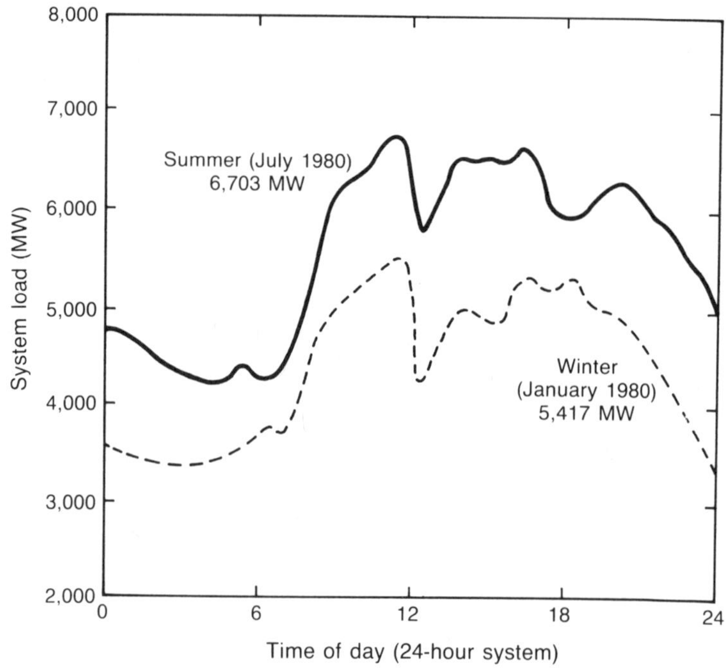

Figure 3.13. Peak winter and summer daily load patterns for Taipower in 1980.

ply occurred entirely on the supply side. Over the long term, this meant that the utility oriented its planning to build the quantity and types of plants needed to meet load. Over the medium term, this meant planning maintenance and refueling to coincide with the appropriate peaks and valleys in load. In the short term, this meant having on-line generating capacity to cover demand fluctuations expected or unanticipated plant outages.

The movement to adopt load management practices recognizes that by integrating the needs and costs of supplier and customers, it is possible to meet electricity needs at a cost that is less for the economy as a whole than is possible with supply modifications alone. It requires, however, a significant degree of additional communication and cooperation between the utility and its customers.

From the standpoint of the general discussion of utility operation above, load management has the potential of improving load factors. Figure 3.16 shows the five actions that load management or demand-side management may achieve. The possibilities include peak reduction, load shifting, valley filling, overall conservation, and strategic actions for building load. Load management methods can be applied to improve load factors through valley

Table 3.3
System Load as Percentage of Annual Peak, 1981

Time Duration (hours)	0	1	2	3
System load as % of annual peak	100	94.1	91.8	90.3
Excess load (%) for 1% excess of hours	—	5.9	2.3	1.5

Table 3.4
Cost of Needle Peak Load, 1981

Item	Cost
Capital cost for gas turbine	= NT$3,022/kW-yr
Energy cost for gas turbine	= NT$4.87/kWh
kWh by excess load (first 1% of hours)	= 0.5 × 6,797 MW × 5.9% × 8,760 hrs × 1%
	= 17.5 x 10^6 kWh
Service cost for excess load	= 3,022 × 6,797 MW × 5.9%/17.5 × 10^6 + 4.87
	= NT$74/kWh

Table 3.5
Industrial Off-Peak Consumption, 1980–1982

	Ratio of Off-Peak to Total kWh (%)	
Item	January–December	January–October
Industries over 500 kW with rising off-peak kWh		
Iron furnaces	51.3	49.0
Plastics	54.3	51.6
Ice-making	47.3	42.3
Sugar mills	44.3	38.1
Nonmetallic mining	46.3	43.2
Textiles	46.6	43.8
Industries less than 500 kW with high 1982 off-peak kWh		
Irrigation pumping	—	76.9
Forms fisheries	—	53.7
Printing	—	48.5
Ice-making	—	46.8
Paper and paper products	—	45.2

Source: Taipower.

— Not available.

filling or simple peak clipping. The appropriate methods and hardware depend on the strategic requirements of the utility.

A load management system is made up of three components. The first is the utility or supplier of electric power. The utility must make the decision as to whether or not it has sufficient energy and/or capacity to sell to meet full demand. The second component is the communications system between the utility and the customer. As will be seen, this communications system can operate over a wide variety of channels and can be one-way or two-way depending upon the requirements of the utility system as a whole. The third component of the load management system is the customer. At this point, the need is for control and for receipt of information from the utility.

It is important to distinguish between decision and control at the customer

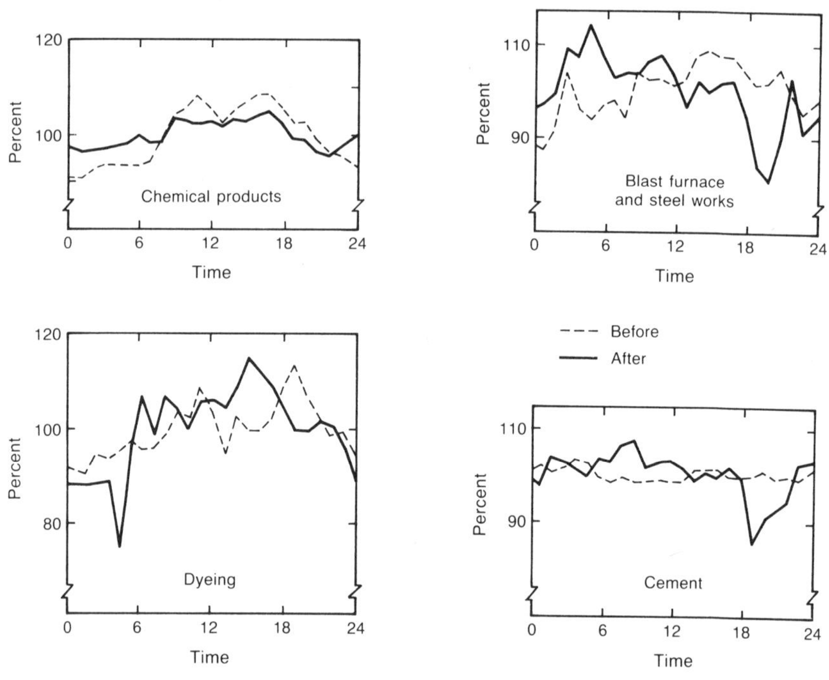

Figure 3.14. Impact of time-of-day tariffs on daily load patterns for six industries in Korea. Most industries responded by shifting loads to off-peak times.

end of the system. In many of the load management systems discussed below, the differentiation between control and decision is not entirely clear. Most customers can easily make decisions as to the criteria they will accept for control of their electric demand. An example might be a willingness of a commercial customer to let refrigerators be turned off for up to an hour once a day if a discount of US$0.01/kWh were offered on his or her bill, but unwillingness if the discount were half of that, for instance. Once the customer has made the decision as to the criteria for control, it is often quite acceptable to the customer to turn over that control to an external system, which may be a computer, an energy management system or company, or the electric utility.

The second major set of issues in load management deals with the frequency of the information flow between the utility and the customer and with the

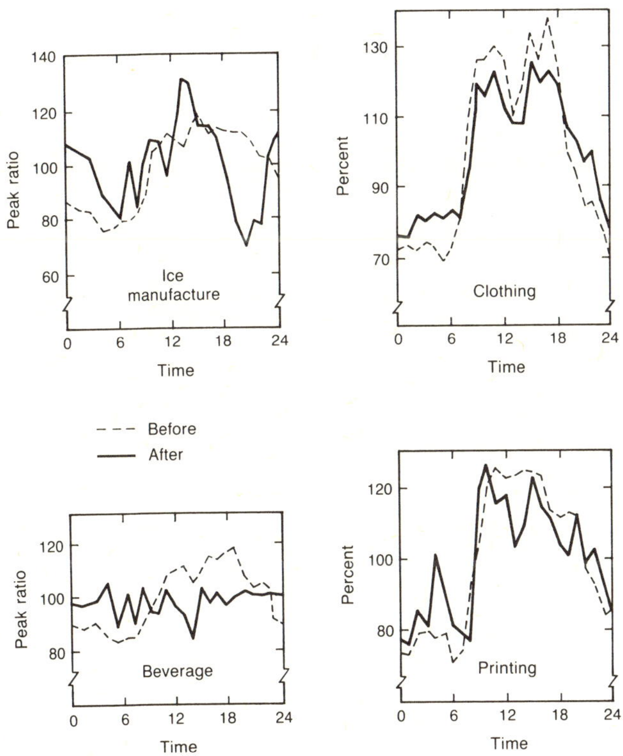

Figure 3.14. *(continued)*

number of levels of information or the number of choice parameters given to the customer by the utility at any given time. Frequently load management is considered in a context of relatively instant or short-term information flow between the utility and the customer, but it need not be. Load management can be considered to include time-of-use rates, which are, after all, an incentive to customers to modify their behavior during peak periods. Thus it is possible that the timing of the changes in information may be as infrequent as once a year, even though the number of levels set—that is, peak versus off-peak, or summer versus winter—may be numerous. As one approaches greater accuracy in expressing actual cost of utility generation to the customer, the number of levels of price or quantity increases, and the frequency of changing the levels may increase as well.

There are also specific issues on the utility side, the customer side, and, in terms of communications, between the utility and the customer. From the utility's perspective, there has been and will continue to be a major dialogue between system operators and load management departments with respect to the type of response required in order to take advantage of the general willingness of customers to have their load managed. To this end, utilities diverge in their opinion as to whether customer relief through load management should be in response to

- Unanticipated system emergencies.
- Times of system peak.
- General needs for leveling load.

Given these three potential areas for response, the information sent by the utility to the customer varies dramatically. If the objective is to obtain relief from unanticipated emergencies from the customer—that is, to gain short-term reliability—then it is necessary for the utility to be able to quickly communicate an on-off (one-zero) type of signal to the customer. The on-off signal must be directed at either large, responsive industrial customers who have contracted for short-term interruptible rates, or must be addressed specifically to appliances that can be taken off-line at times of system emergency. The tradition of water heater controls and interruptible services to customers with large electrical loads falls into this category. These latter appliances may be spread more evenly through the utility system in, for instance, residential facilities. To fulfill this expectation requires very rapid response time on the part of the load management structure.

Another type of signal for short-term requirements is currently in the experimental stage of development in the United States. It is an on-off signal, but one that reduces load to a demand level that residential customers contract to beforehand. In the Southern California Edison service territory, engineers are evaluating the potential for the demand subscription service (DSS) in which residential customers are to limit their demand—that is, their kW load—to a specific level when requested by the utility. Under these circumstances, an on-off signal is sent from the utility to the customer stating that the customer must reduce the load within a certain amount of time. If the load is not reduced below the contracted level, then *all* of the customer's power is shut off. The customer does have the right to reset the system and to reduce the load to meet the demand limitations for which the customer previously contracted.

The second category is response at times of system peak, particularly when the marginal operating cost is extremely high, such as those times when gas turbines are being operated. Under these circumstances, the utility once again wants to send a reasonably quick signal to customers to have them respond both at a time and in a manner that is beneficial to the utility. Air-conditioning cycling systems offer one such mechanism to many utilities. Again, the signal is an on-off signal to air-conditioning control units; those units in turn cycle air conditioners (20 minutes on, 20 minutes off, etc.). A "softer" version of this response pattern can be seen in flexible time-of-use rates, in spot pricing rates, and in other time-differentiated but continuous rather than discrete signals sent from the utility to its customers.

Within the industrial sector there are other load management techniques available that can be used to influence customer behavior. All of these methods give the customer a clear benefit, but all also involve a utility benefit. These methods generally fall under the category of interruptible rates, which are rates that give the utility the right to demand a specific response from the customer upon call. The mechanism for customer contact can vary from a simple telephone call to communication through an elaborate communications system. The action can vary from a complete plant shutdown to curtailment of only a specific amount of the load. The length of the response required can be from only a few minutes to several hours. In general this type of load management is in response to more prolonged peak relief requirements rather than to emergency control requirements. It is generally aimed at customers that have either a long-term "storage" type of load or at customers who are particularly sensitive to prices of energy. Under these circumstances there is either a small regret function associated with the loss of electricity for a specific period of time or a high reward in the form of lower electricity costs for the willingness to reduce the load below a level agreed upon in advance.

Energy storage by the customer can often be quite cost effective compared with supply management by the utility. Energy storage can be accomplished by a wide range of actions by different customers. For residential and commercial customers, for example, refrigerators and hot water heaters can be modified to increase their intrinsic storage times. Specific cost-effective systems for cold storage are available for nighttime filling in commercial buildings (McCannon, 1983). For industries with specialized electricity-intensive process steps, for example, energy storage could consist of bins of parts, that is, electricity-intensive parts would be made at times of low load. Clearly, therefore, load management through energy storage may involve restructuring production and inventory strategies as well as installing storage equipment.

The third type of response that utilities seek is one of general flattening of their load curve by encouragement of increased consumption during periods of excess capacity, frequently at night, and discouragement of additional consumption during peak times in the late afternoon. Under these circumstances the speed of communication need not be great. Irrigation pumping is an example of this type of load management that can be cost effective for everyone involved (Schneider, 1983).

In summary, the nature of the signal and action sent from the utility to the customer is a function of the type of response required and of the purpose for which the response is required. Some utility operators have argued that instantaneous positive responses from customers are the only type of value to them. It appears, however, that in many instances smaller responses on the part of customers can have a high savings value to the utility. In addition, the response does not have to be uniform across all customers as long as a predictable percentage responds (i.e., a stochastic response).

On the customer side, the response to signals sent from the utility can once again be divided into a number of categories. The rationale for response, however, is one that is generally felt to be determined by the monetary savings associated with the action and its discomfort and inconvenience. It has been shown that with no change in customer requirements or customer comfort, an enthusiastic response to load management is possible. Residential and commercial customers, in general, do not want to give up personal comfort for a small savings. In the industrial sector, however, it has been shown that significant savings can be had both by the utility and by the customer. Customers can either have their own control systems that perform a specific logic each time a utility signal is sent down, or they can have the utility signals received as on-off variables for individual appliances or process lines within a specific factory. In this way, customers can respond to either a discrete, on or off signal from utilities, or they can respond to continuous signals. Continuous signals can be expressed either in quantity terms, indicating a limit of demand during certain periods, or in price terms, indicating an increase in tariffs for demand above certain levels.

Illustrating this process may be valuable. Through a price incentive a customer can decide in advance to be willing to have specific functions in his industrial facility turned off at times of system emergency. He can then establish a system within his own plant such that when the signal comes from the utility, someone or some machine adjusts the plant equipment according to the preplanned response.

The objective of load management from the customer's perspective is generally to have lower cost while maintaining service quality. Although this objective seems difficult to meet, it is clearly not impossible given that many customer devices that have ''storage'' or rescheduling capabilities are available for load control. Such devices and processes as electrical furnaces, large-scale water heating systems, cooling equipment for large commercial buildings, pumping water, and compressing gases offer systems that can operate as low-cost storage for electric power even though they cannot actually generate electricity. It is possible to see, when focusing only on these systems, that the utility could control the load during short-term emergencies without having any noticeable impact on the customer's quality of service.

Industrial customers are generally more responsive to load management implemented by economic signals than are commercial customers. Commercial customers are, however, more responsive than residential customers. The quality of response is generally based on the extent and clarity of economic signals from the utility to the customer and vice-versa and, significantly, upon the size of the load that can be managed.

The response that can be expected from the customer can take one of four forms. The first is that the customer does not respond in any way. The signal is received, but the customer does not act. The second is that upon receiving a signal from the utility the customer acts manually. An individual makes a decision and that individual turns off appliances and resets thermostats. The third is an automatic response mode. The signal is received from the utility and appliances are turned off at the customer end automatically. The fourth mode of response is likely to have the largest effect in the future—an intelligent response. In this mode, customers will receive within their facilities the signal for action, which will then be filtered through an intelligent decision and control system that will maximize the benefit to the customer (in terms of the cost of electricity) and minimize the impact on the customers' operating conditions within their own facilities.

Physical Communications Systems

The design of the physical communications system between the utility and the customer is open to considerable flexibility. At its most basic level, no communications are required between the utility and the customer except for an initial agreement at the beginning of a contract term. In the early days of load management, time clocks were used to turn hot water heaters on and off such that they were not on during periods of peak demand. ''Communication'' in such a system consists only of the monthly bill and an occasional visit to reset the clock.

A second form of communication that requires no specific hardware is that of public appeals for energy conservation and/or public appeals for capacity relief at times when the utility system is under extreme stress. It has been shown that such systems as radio appeals will, for a limited number of days, offer the utility significant peak relief.

The third communication medium between the utility and the customer, although noninstantaneous, may be so simple a system as the publication of short-term prices for the next 24 hours in the morning newspaper. Such systems have the advantage of requiring no dedicated communications systems.

Within the literature on load management there are four distinct categories of communications systems (Talukdar and Gellings, 1986). The first is power line carriers, which breaks into two subcategories: those associated with modulation of the 50 or 60 Hz wave form, and those that are operated through a ripple control system. In general, the power line carrier systems operate best

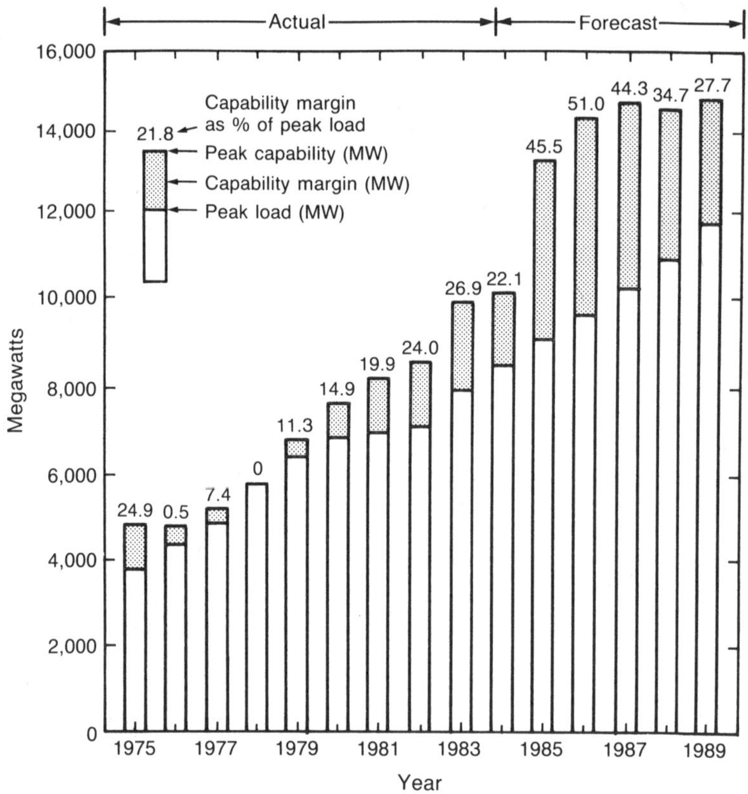

Figure 3.15. Historical and forecasted trends of peak capacity, peak load, and capability margin for Taipower. Note low margins in the late 1970s.

in off-on control environments. In tests they have shown themselves to be unreliable in some environments. Problems owe largely to the difficulties of moving either ripple or wave modifications through transformers in the power system (Flory and Parker, 1985).

The second major mode of transmission of information to customers from the utility is existing telephone lines. Telephone lines can be operated in a one-way or two-way mode and can thereby be used for billing and meter-reading as well as information transmission from utility to customer. Telephone lines have the advantages of being relatively clean, having a two-way communications capability, and, in many areas, excess capacity. Therefore, low marginal costs are possible, provided the service is not extended to the point of competition with the current phone systems. Telephone systems have a disadvantage from the perspective of the electric utility of being owned by another utility and thereby regulated outside the electric utility environment. Without control of the wires, many electric power companies feel that they

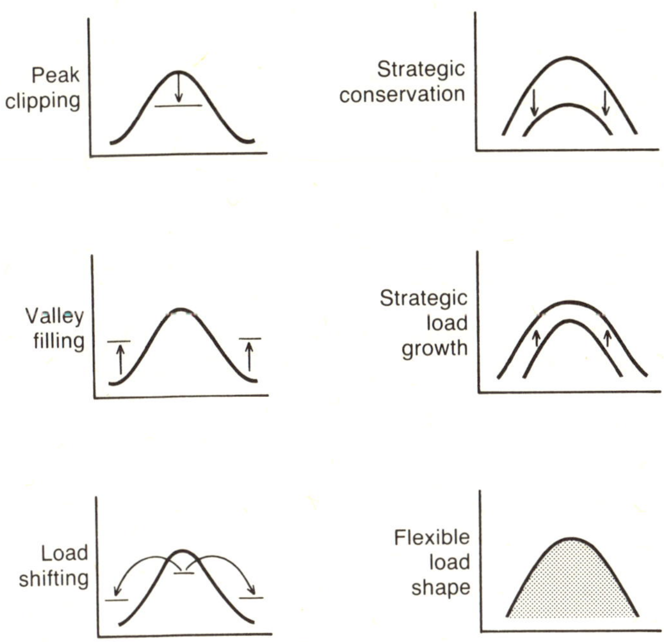

Figure 3.16. Illustrative effects of different load management tactics on daily load patterns.

are at the mercy of the telephone utilities or their regulatory bodies in terms of rate-setting and in decisions about peak versus off-peak usage.

The third type of communications system between the utility and the customer is the use of coaxial cable or cable television systems for broad-band communication from a central point to a customer. Broad-band communication is necessary for sending large amounts of information (e.g., to control a range of devices to different degrees). Such broad-band communications systems, although allowing for two-way communications, are more ideally suited to one-way communications from the central point to a range of receiving units.

The fourth type of communications device is a radio signal either dedicated in the short-wave band lengths or in an oversignal on a commercial radio system, both of which are best operated on a one-way rather than a two-way basis. There is currently available a set of novel combinations of both telephone and broad-band cable systems for communication between the utility and the customer. These hybrid systems are still being developed, and are expected to be commercially available in the late 1980s. Such systems have the advantage of being able to utilize broad-band communications in an outflow direction from the utility to the customer and using dedicated narrow-band communications for data pickup from metering devices.

There is one type of instantaneous load management technique that is used in many instances: lowering voltage or frequency. Such an action does temporarily reduce load, but can also damage power-consuming equipment. Thus it is not considered to be anything other than an emergency measure.

Communications devices should be chosen based on the requirements of the particular utility and customer. Whereas one could not argue that it would be possible to have a total menu of options within any given utility district for communication between the utility and thc customer and reverse flow communications from the customer to the utility, it is possible that more than one, or even more than one hybrid, could be available within any service territory.

In summary, the future of load management systems that are physically based will undoubtedly be in the area of greater intelligence or logic within both the utility system and within the customer system. The prices of microprocessors and communications equipment are dropping. As a result, the capabilities of load management systems to simultaneously fine-tune the customer load as well as the utility supply will grow.

Given this prospect, in the future there will undoubtedly be greater customer differentiation with respect to load management. The targets of opportunity within customer loads will be easily identified and used by the utility to its advantage as well as to the customer's. It is thus expected that in the future both the utility and the customer can significantly profit from this fine-tuning.

Financial Incentives

In the most general sense, the objective of the regulated tariff setting is to recover, over the length of all investments, the full amount of investment capital, all operating costs, and a return on capital investment. In addition, tariffs are to be fair in that they will reflect the actual operating costs to the individual customer classes. They should also be consistent in having a minimal amount of cost subsidization between individual customers within a class and between the customer classes themselves.

Given the objectives, however, it is possible that if the price charged to customers is to reflect the utilities' costs as implied above, an element of time and possibly also of geography should be introduced into the tariff structure.

The literature on tariff structures is complex and based in large part on a combination of economic theory and practical application. As discussed in Chapter 6, the philosophy in tariff structures is generally divided into two groups, one that looks at tariffs being set by average costs and the other in terms of marginal costs. In the past, when the fixed cost component of the tariff (i.e., the capital component) represented between two-thirds and three-quarters of the tariff rate, average costing principles offered a reasonable approach to choosing the tariff rate. At present, however, fuel costs, or short-run marginal costs, represent between 50 and 60 percent of total supply costs. As a result, short-term marginal costing has become a more important component and therefore a more dominant practice in the setting of tariffs. In addition, the increasing incremental costs associated with capacity additions have made the philosophy of long-term marginal costing equally important in the setting of tariffs (Munasinghe and Warford, 1982). Spot pricing, which is based in large part on short-term marginal costing, may offer one solution to both the setting of tariffs in general and the reconciliation of alternative tariff structures to individual customers (Berrie et al., 1982; Caramanis et al., 1983; Schweppe et al., 1988).

Tariff structures have, traditionally, been made up of two components. The first component is an energy cost reflected in a kilowatt-hour cost based on usage. The second is a demand component based on the kilowatt demand of an individual customer. In general it is only the commercial and industrial customers who face two-part tariffs, but there is a long-established tradition in Europe and a recent trend in the United States for the larger residential customers also to face a two-part tariff.

Block rates may be either increasing (as one consumes additional energy) or decreasing. In the past, many utilities have looked at decreasing block rates in an effort to encourage consumption. At present, in the United States and in Japan increasing block rates are applied to a number of customer classes.

Seasonally adjusted rates reflect the summer and winter cost differential within a utility system. These rates are becoming more and more common and reflect a flat rate structure but with a seasonal adjustment.

Since the principal goal of load management is to equalize the marginal cost of supply and the marginal benefit of demand, financial signals are an integral part of the necessary communication between supplier and customer. There are a number of forms that are possible for these signals.

Time-of-use rates reflect a time differentiation between peak and off-peak hours, weekdays and weekends, and summers and winters. This offers a way either annually or seasonally to adjust rates to reflect (on average) the time-variant nature of the cost of generation. Spot pricing represents daily fine-tuning relative to time-of-use and seasonal rates and reflects with a high degree of accuracy the real-time cost of generating electricity. In addition, spot pricing also offers a way of rationalizing the calculation of all other rate structures including flat and seasonally adjusted rates.

Interruptible rates are price-quantity contracts in which the utility lowers a customer rate based on that customer's willingness to be interrupted on short or long notice for what may be either a short (minutes) or an intermediate (hours) period. They have been in use for a long time in some European countries.

Demand limiting tariffs are currently under development in the United States and Western Europe as well as in Northeast Asia. These tariffs reflect an agreement on the part of the customer not to exceed a specific demand (measured in kilowatts) level during either an annual period or, as is more frequent in the United States, during periods of system emergency.

Table 3.6 summarizes the relative effectiveness of using price and quantity load management techniques to achieve several important utility objectives. If the objective is to deal with unanticipated system emergencies, then quantity-based systems are more effective than price-based systems. If the objective is to deal with times of system peak, then the two are roughly equivalent. If the objective is to flatten the load, then price incentives are probably more effective than quantity-based systems.

Financial incentives are becoming more attractive as load management strategies as the flexibility of these techniques is being recognized by utilities. The closer the financial signals can match the real-time costs of the utility,

Table 3.6
Effectiveness of Price and Quantity Load Management Techniques

Objectives	Price	Quantity
Handling unanticipated system emergencies	+	+ + +
Managing times of system peak	+ +	+ +
Encouraging general load smoothing	+ + +	+

+ + + Most effective

+ Least effective

the greater the flexibility. There a number of examples of real-time rates in the United States and Europe, and some in Japan. France has an estimated 50,000 industrial customers on real-time rates, apparently with success (Meta Systems, 1987). The U.S. rates currently in place range from residential experimental rates to fully implemented industrial programs. The rates are currently receiving additional attention as utilities are being pushed to justify and rationalize all tariff structures across all customer classes.

WHAT THE FUTURE MAY BRING

The evolving tool of load management has only just started to find a role in Northeast Asia. There would seem to be many applications of load management techniques that can assist the utilities in the region in handling their growing problems with load factors. In addition, over the longer term, load management tools may have a significant impact on the pattern of power system growth in other developing countries, which may in turn be able to learn from the experience of Northeast Asia.

Short-term Application in Northeast Asia

Earlier we outlined the problems being experienced in Northeast Asia in matching load and supply and briefly described the present way in which load management techniques have been applied. Some of the ways in which load management can be strengthened are addressed below.

The future trend of the industrial structure forecast for Japan has generation-cost implications that are not compensated for by the characteristics of the hydro-power system, which lacks the capability to store the major spring river flows. There is only a factor of 2.5 difference between the high and low outputs from the hydro stations, a situation that is unlikely to change because of intrinsic geographical conditions at the hydro sites. Furthermore, owing to cost and site availability, most of the pumped storage that is now planned and under construction is such that upper reservoir capacity is designed exclusively for daily regulation. Given this set of conditions in Japan, the use of summer and winter rates is easily justified. TEPCO's daily load curve also exhibits a large potential for time-of-use rates. This is particularly so given the large size of some of its industrial customers.

The present rate structure in Japan and the present programs of the utilities are focused, in part, on increased load shaping. Many of the special rate structures for industrial customers explicitly take into account the cost savings advantages of load shifting. In addition, efforts under way to promulgate appliance efficiency standards for air-conditioning and refrigeration form an important component of a short-term strategy for load management in Japan.

Both Korea and China (Taipei) face similar short-term problems. Evolving industrial development and increased personal income are leading to reduced utility load factors. In both countries, the more general trends have been exacerbated by generally slower economic growth and growth in electrical demand. All of these forces point to the short-run potential of demand-side load shaping and/or load control. The efforts required and the potential benefits extend well beyond the electric power sector. Direct governmental involvement may be required.

As an example, the daily load curves in all three countries show a large demand valley between noon and 1 p.m.—what might be called the Northeast Asian lunchtime slump. This dip in demand is operationally difficult and expensive for the utility. Simply staggering the regular business hours of industrial and commercial firms would alleviate a portion of the slump. At the same time it might reduce the burden on the transportation system during its own peak hours, decreasing the capital requirements of that sector. Other utility incentives such as special rates may also fill this valley.

Flexible working hours were tried in Korea some years ago, affecting both the industrial and the commercial sectors, but the measure encountered strong popular opposition. Similarly, the Christian community strongly opposed an attempt to more evenly distribute work hours among the seven days of the week. Economic incentives may be more effective than legislative action under these circumstances.

Several nonexclusive options seem to be available to both KEPCO and Taipower for load management. Time-differentiated rates may be extended to all classes of customers where the benefits exceed the costs of metering and billing. Such rates encourage increased off-peak consumption and may provide added (nonlegislated) incentives for load shifting through process change or through storage of either product or energy. Pricing structures offer indirect control options. More direct actions can be taken through interruptible contracts or, under specific circumstances, through demand limitation or appliance cycling. All such actions require dedicated signaling, control, and metering equipment. These reflect costs that must be weighed against the benefits gained from load management.

A second approach would be direct control of the peak-load-consumption devices themselves. Taipower is currently conducting the research required to identify those residential and commercial appliances. At the industrial levels, direct control frequently implies the use of interruptible service contracts. Under such contracts, large customers contract for a demand (kW) limit based on a prespecified warning time and duration of interruption. Under these circumstances, the customer chooses the amount of load to be placed on an interruptible rate and, at the time of receipt of warning, the precise load to be shed. Such an approach provides occasional peak relief (possibly 20 times per year), but it does little to fill the off-peak valleys.

Energy performance standards for buildings offer yet another distinct approach. Although significant for control of temperature-sensitive loads in the long run, it has no effect on short-run load management problems. Improving building standards requires a legislative mandate followed by creation of a bureaucratic operating structure. Economic incentives frequently provide the same end effects at a far lower implementation cost. Standards for residential and commercial equipment are more effective because consumer awareness of the life cycle cost is generally not sufficient to overcome the first cost bias. It is well to remember that appliances such as residential and commercial air conditioners contribute significantly to the weather-sensitive peak of both KEPCO and Taipower.

Government actions often adversely affect the utility load distribution. Programs for industrial development can add significantly to peaking problems for the utility. Other actions have caused major problems in distribution of load (load factor). Programs that focus on single shift operations move load to the peak periods. Over time this need not be bad given some amount of planning for a utility system that is not concentrated on large, inflexible baseload capacity, as is the case at present. In addition, government policy such as that in China (Taipei), which has encouraged reindustrialization away from industrial facilities that create base demand, tends to exacerbate any problems that already exist in load balance. There is clearly no correct answer to the question of the best capacity mix for a particular utility.

Most utilities and government planning groups are realizing that flexible and robust systems are best for adapting to uncertainty. A good solution is one that provides the basis for an out-of-balance capital stock to evolve to a more balanced stock as demand grows. The evolution or change can be encouraged by load management programs, particularly those based on financial/economic incentives, because these tend to be self-correcting, that is, they work toward equilibrium. In addition, significant benefit comes from the awareness on the part of governments, customers, and utilities of the need for demand-side as well as supply-side programs to smooth the impact of demand changes on system reliability (Tabors and Schweppe, 1984).

The rest of the century will be a new era for the utilities of the developed world and, to a growing degree, those of the NICs such as Korea and China (Taipei). Conservation and competition will dominate (*Electrical World,* 7/84, pp. 69–71). Repowering old plants rather than building new ones may well be more cost effective. Fitting the supply to the load and improving the cost efficiency of loads will be the parts of the business that will grow. Unless utilities begin to take on these unfamiliar roles, they will find themselves shut out of much of this growth.

Long-term Implications for Developing Countries

Although many of the issues are country- and system-specific, much can be learned about load management from the experience of the United States, Western Europe, Japan, Korea, and China (Taipei) that would be of value to developing countries. These lessons are both positive and negative. The western electric power systems are rigid in their structure and operation. They have a low turnover in capital stock and, in many cases, have excess capacity. Ironically, although such characteristics are partly a result of a lack of attention to load management in the past, they tend to reinforce a persistent lack of interest. In addition, these systems are largely based on capital built before the dramatic change in relative energy prices, leaving them with relatively suboptimal overall plant mixes. Utilities that are developing rapidly have the opportunity to choose a different mix of plants, a different structure for tariffs, and a different system for management of loads.

Some observers may believe that it will be possible for both developed and developing power systems to follow old pathways of planning, operations, and management. This seems unlikely because the economics of electric power have changed dramatically in the past decade and will not return to their previous state. Capital cost for new capacity has increased dramatically. This increase has meant that many of the old rules of thumb of system composition and operations no longer hold. In addition, both fuel costs and uncertainties about them increased dramatically; costs *may* either increase or decrease significantly in the future. This again has changed the relative advantage of generation and such alternative technology as load management. Finally, customers' willingness to tolerate price uncertainty in electricity as well as general price increases in electricity has changed.

In addition to economic considerations, three physical issues have dramatically changed the environment in which electric power systems operate. First, there is now available a set of new energy generation technologies that range from better, more efficient use of fossil and nuclear fuels to the economic acceptability of renewable energy technologies (see Chapter 4). Second, as described above, there are new technologies for intelligent customer and utility control. This may have a significant effect in the future on utility operating practices and customer response. Third, there are new technologies currently available and on the drawing boards that will make communications between utilities and their customers more rapid, more efficient, and less expensive.

Given the rejection of the old path, what lies ahead? First, in developed electric power systems, the future will focus on greater and more efficient use of existing generation facilities. This implies an emphasis on load control and load management in order to balance the demand side of the equation with the supply side. Second, it implies that there will be greater retrofitting of existing generating plants to make them more fuel efficient and more responsive to the needs of the utility system and to extend their productive lifetimes. Third, there will be larger and more frequent real-time information and power flows between the individual utilities in a system in order to benefit from the relative advantages in generating cost between locations within countries and between utilities.

Focusing on the rapidly developing utility systems presents a totally different picture. Additional generating options will soon be available for consideration. They will probably have centralized technologies that could be larger in scale than those currently available or, more likely, smaller in scale than those currently in use. In addition it will be possible to look at decentralized generation on an equal footing with centralized generation. Rural electrification, for example, may be accomplished more quickly and at lower cost in many areas by incorporating small-scale generating units into the system rather than by extension of the central grid. Alternative highly flexible, reasonably reliable local *power systems* are becoming available. In many of the developing power systems there will be the opportunity to look at electronic rather than mechanical control systems for the utility. The revolution in microprocessors is bringing about a major revolution in the control of utility substations, switchyards, and even power plants, allowing them to operate more efficiently and automatically.

Throughout the utility system, it will now be possible to have more greatly distributed intelligence, which will allow for both more secure systems and softer systems in terms of the system control logic.

Given the growing possibility of increased flexibility in power systems, three trends emerge. First, it will be possible to have independent nodes of generation, implying that reliability of the node will match the needs of the consuming public at that node. When increased reliability is required and when it is economically feasible to hook up to a central grid, the generating units can either be used in backup form or they can be moved and used somewhere else.

Second, it will be possible in the developing system to have weaker links between the nodes once they are interconnected. The logic of this follows both from the increased intelligence within the system in terms of automatic control and switching and also from a desire to take advantage of the economics of lower reliability in areas not able to afford a "one day in 10 years" LOLP standard.

Third, within the developing systems as within the developed systems, the ability to communicate between the customer and the utility will allow those systems to develop with lower spinning reserve margin requirements and, quite possibly, lower overall reserve margins. Reserves can also be provided on the demand side rather than only on the supply side.

The implications for the future of electric power development may be expressed in two ways. The Western or developed model for power system growth and operations may no longer be applicable in the context of rapidly developing power systems. The rigidity of the developed systems of today, their inability to incorporate new technologies because of low growth in demand, and their relatively rigid reliability standards make them less capable of change and adaptation than the rapidly developing systems.

The second point is that the developing systems may, because of the above discussion, become the leaders rather than the followers in developing new patterns of utility operations, management, and control.

To realize this future, however, will require that the developing power systems look at the recent past to evaluate changes in control, generation, and management, and to project this recent past into the future to determine the direction they may want to pursue economically. To do this will require an increase in human resources—human resources able to accept the challenge of these new possibilities. Utilities should avoid becoming locked into large-scale generation and transmission without considering alternatives in source, scale, and operations.

Finally, it will be necessary to convince the funding sources, be they public, private, domestic, or international, that alternative structures for power systems may be cost effective. For this to occur, funders must perceive the potential economic advantages of these alternatives and be able to weigh these risks against the likely returns.

NOTES

1. Load factor alone is not a sufficient indicator of system efficiency. It is possible, for instance, for one utility's load factor to be lower than that of a second utility, yet the first utility may operate more efficiently because its mix of plants was chosen to minimize total costs based on known patterns of demand. This same caveat may hold for comparisons of different time periods for a given utility. We use the comparison correctly in this chapter because all three utilities are actively strengthening their baseload. This change reflects a desire for an improved (higher) load factor.

2. The same warning as in note 1 concerning cross-national comparisons may hold for intertemporal comparisons of a single utility.

3. Load management is now frequently referred to as demand side management or load shaping. This accounts for the need both to reduce peak loads and, in many utilities in the United States today, to increase all system sales.

4. Increasing block tariffs provide limited marginal production cost information to the customer. Consumption levels by billing period are divided into fixed blocks of, for instance, 100 kWh. Price per kWh increases as one's consumption level moves from one block to the next. Declining block rates are the exact reverse; subsequent blocks are less costly.

REFERENCES

Berrie, T. W., Mallalieu, B. D., and Mylon, K. R. D. (1982). "Interactive Load Control and Energy Management." Annual conference of International Association of Energy Economists, Cambridge, MA.

Caramanis, M., Schweppe, F. C., and Tabors, R. D. (1983). "Spot Pricing and Its Relation to Other Load Management Methods," MIT-EL 83-001. Cambridge, MA: MIT Energy Lab.

Electrical World. Issue and page references in text.

Flory, J., and Parker, O. L. (1985). "Communications and metering equipment for spot pricing." *IEEE Power Apparatus and Systems* 104(6):1363–1369.

McCannon, L. (1983). "Thermal Energy Storage for Commercial Space Cooling." San Diego Gas and Electric Co.

Meta Systems, Inc. (1987). "Real World Experience with Real Time Pricing." Report to the Office of Coal Policy, Department of Energy. Washington, D.C.

Morgan, M. G., and Talukdar, S. N. (1979). "Electric Power Load Management: Some Technical, Economic, Regulatory and Social Issues." *Proceedings of the IEEE* 67(2):241–312.

Munasinghe, M. (1979). *The Economics of Power System Reliability and Planning*. Baltimore: Johns Hopkins University Press.

Munasinghe, M., and Warford, J. J. (1982). *Electricity Pricing*. Baltimore: Johns Hopkins University Press for the World Bank.

P.G. & E. (1983). "Load Management: 1984. Test Year." Pub. No. 12. San Francisco, CA: Pacific Gas and Electric Co.

Schneider, S. (1983). "1982 Preliminary Analysis Results of the Agricultural Load Management Program," PA-2X. San Francisco, CA: Pacific Gas and Electric Company.

Schweppe, F. C., Caramanis, M. C., Tabors, R. D., et al. (1988). *Spot Pricing of Electricity*. Norwell, MA: Kluwer Publishing.

Tabors, R. (1985). "Recent Research in Electric Power Pricing and Load Management," Energy Research Group Monograph #85. Ottawa and Tokyo: International Development Research Centre and United Nations University.

Tabors, R. D., and Schweppe, F. C. (1984). "Electricity spot pricing in developing countries." *National Development Magazine* (Nov./Dec.).

Talukdar, S., and Gellings, C., eds. (1986). *Load Management*. New York: IEEE Press.

CHAPTER 4

PATTERNS AND IMPORTANT ISSUES IN SUPPLY

Y. H. Kim and P. S. Pan

Planning for the supply of electricity involves not only the obvious, such as a knowledge of primary energy supplies and generation technologies, but also, and more critical in many cases, an understanding of the implications of uncertainty and close attention to management. With this in mind, this chapter reviews the growth of power systems in Northeast Asia, noting particularly how they have changed during the economic development process, addresses some of the more general issues of system planning, and ends with instructive observations based on that history.

HISTORICAL GROWTH AND TRANSITION

The levels of economic development reached by China (Taipei), Japan, and Korea are widely different, and this is reflected in their electric power statistics. As noted in earlier chapters, in 1980 Japan had a per capita electricity consumption of 4,450 kWh, China (Taipei) 2,330 kWh, and Korea 860 kWh. Similarly, the total installed generating capacity of Japanese utilities was 129,360 MW in 1980, about 14 times greater than either that of the Korea Electric Power Corp. (KEPCO) (9,390 MW) or Taipower (9,060 MW).

In contrast to their relative sizes, all of these power systems share a common history of expansion at high rates. As shown in Table 4.1, from 1950 to 1980, Japan's installed capacity grew from 9,050 MW by a factor of 14, an average annual increase of 9.3 percent. Over the same time period in China (Taipei) the installed generating capacity increased from 276 MW by a factor of 33, an average annual growth rate of 12 percent. The installed generating capacity of Korea grew at an average annual rate of 13 percent during the same time period, or a factor of 41 over the base at 231 MW. Table 4.1 also breaks down installed generating capacity by plant categories for selected years.

As Table 4.1 indicates, each of the utility systems relied on hydropower in the early years. Hydro was attractive because it was both locally available and economic, and low growth rates made possible reliance on these relatively long-term projects. Once a high growth rate was achieved, however, generation additions were diversified into fossil fuels—oil, coal, and natural gas. Although these could be implemented at the speed and quantity needed, they were also imported. Consequently, there has been a shift from hydro-dominated to thermal-dominated systems since 1962 in China (Taipei), 1961 in Japan, and 1957 in Korea. This transition represents the typical development pattern of a growing power system in energy-deficient countries.

As Table 4.2 indicates, Japanese utilities and Taipower (and KEPCO at times) experienced inadequate capacity reserves during 1960–1980. Although

Table 4.1
Installed Capacity by Type of Plants, 1950–1980

	1950	1955	1960	1965	1970	1975	1980
Japanese utilities							
System total (MW)	9,046	12,190	20,650	36,500	58,960	99,740	129,360
Source/type (%)							
Oil	—	1.3	11.8	24.8	40.1	59.1	46.1
Coal	—	32.7	31.2	32.0	23.5	5.7	4.1
Nuclear	NA	NA	NA	NA	2.2	6.5	12.0
Hydro	67.8	65.5	56.7	39.6	26.3	16.7	13.9
Pumped hydro	NA	0.4	0.3	2.2	5.8	7.1	8.4
Natural gas	NA	NA	NA	1.4	2.0	4.7	15.2
KEPCO							
System total (MW)	231	301	367	769	2,508	4,720	9,391
Source/type (%)							
Oil	19.9	21.3	0.3	9.0	65.5	72.0	73.4
Coal	53.2	40.9	60.7	73.1	21.4	14.8	8.0
Nuclear	NA	NA	NA	NA	NA	NA	6.3
Hydro	26.8	37.8	39.0	27.9	13.1	13.2	8.0
Pumped hydro	NA	NA	NA	NA	NA	NA	4.3
Taipower							
System total (MW)	276	493	709	1,186	2,720	5,300	9,056
Source/type (%)							
Oil	NA	8.1	1.0	9.9	26.6	53.6	59.8
Coal	19.8	20.7	35.8	32.4	36.0	18.5	10.8
Nuclear	NA	NA	NA	NA	NA	NA	14.1
Hydro	80.2	71.2	63.2	53.0	33.1	25.7	15.3
Pumped hydro	NA	NA	NA	NA	NA	NA	NA
Natural gas	NA	NA	NA	4.7	4.3	2.2	NA

Sources: Ke et al. (1983); Kim et al. (1983); Kimura et al. (1983).

— Not available.

NA Not applicable.

the utilities might have found it advantageous to have the high plant factors as well as the probable lower production costs that accompany them, serious concern was expressed from other sectors about the adverse effects on government efforts to achieve rapid industrialization. A steadily increasing and reliable supply of electricity is important for the success of such efforts.

Capacity and generation. By 1965 hydropower's share in total generation had significantly dropped in all three countries. Continuous decline in the share of hydropower was balanced by rises in the share of first coal and then oil. In an attempt to phase out oil as a dominant fuel for generation, all three systems have opted for nuclear power. Consequently, nuclear generation has substantially increased, starting in 1970 in Japan and 1977 in the other two countries. In 1980 the share of nuclear generation had increased to 9 percent in Korea, 16 percent in Japan, and 19 percent in China (Taipei). Natural gas in Japan has also shown a sharp increase, rising from 0.5 percent of total generation in 1965 to approximately 18.0 percent in 1980. By 1978 natural gas was phased out as a primary fuel by Taipower, and in 1980 the electricity generated by coal-fired units increased sharply to 14 percent of total generation (see Table 4.3).

As Table 4.3 indicates, in 1980 hydro and gas-fired units together accounted for approximately 35 percent of total generation in Japan. Japan, therefore, has a flexible mix in terms of load-picking potential in times of system stress (see Chapter 3). On the other hand, Table 4.4 indicates a decline in the proportion of generation accounted for by both cycling and peaking units in the three systems. This is apparently in response to the need to conserve fuel.

It is apparent from the plant capacity data in Table 4.5 that the electric power supply systems have a reasonably flexible generation mix under de-

Table 4.2
System Gross Reserve Rate by Percentage

Year	China (Taipei)	Japan	Korea
1950	52.7	—	—
1955	46.7	30.6	109.0
1960	11.6	32.2	27.4
1965	11.2	35.9	27.7
1970	27.6	20.4	61.3
1975	40.8	34.6	40.9
1980	35.1	45.2	75.4

Sources: Ke et al. (1983); Kim et al. (1983); Kimura et al. (1983).

Note: $\text{gross reserve rate} = \dfrac{\text{(system installed capacity)} - \text{(peak load)}}{\text{peak load}} \times 100$

— Not available.

Table 4.3
Net Generation by Energy Sources, 1950–1980

	1950	1955	1960	1965	1970	1977	1980
Japanese utilities							
System total (10^3 GWh)	40.3	54.6	101.7	167.7	307.5	468.0	514.1
Source/type (%)							
Oil	NA	1.5	16.7	31.6	57.9	63.7	44.8
Coal	NA	19.8	31.1	26.1	13.1	3.9	4.4
Nuclear	0	0	0	0	1.5	6.8	16.0
Hydro and pumped hydro	82.1	78.7	52.2	41.8	23.9	15.1	16.7
Natural gas	0	0	0	0.5	3.6	10.5	18.1
Industrial cogeneration	0	0	0	0	0	0	0
KEPCO							
System total (10^3 GWh)	0.382	0.835	1.609	3.050	8.780	26.073	36.047
Source/type (%)							
Oil	14.9	24.2	NA	4.2	57.6	86.3	76.7
Coal	60.7	19.4	64.1	72.6	28.5	5.0	6.4
Nuclear	0	0	0	0	0	0.1	9.0
Hydro and pumped hydro	24.4	56.4	35.9	23.2	13.9	5.2	5.2
Natural gas	0	0	0	0	0	0	0
Industrial cogeneration	—	—	—	—	—	3.4	2.7
Taipower							
System total (10^3 GWh)	1.040	1.966	3.628	6.455	13.213	29.724	40.813
Source/type (%)							
Oil	NA	8.0	0.9	13.4	69.0	79.7	59.6
Coal	6.6	14.1	42.2	44.4	5.6	5.4	14.2
Nuclear	0	0	0	0	0	0.3	19.1
Hydro and pumped hydro	93.4	77.9	56.9	40.1	21.5	13.5	7.1
Natural gas	0	0	0	2.1	3.9	1.1	—
Industrial cogeneration	—	—	—	—	—	—	—

Sources: Ke et al. (1983); Kim et al. (1983); Kimura et al. (1983).

NA Not applicable. — Not available.

mand uncertainty. This is so because the oil-fired and gas-fired units are presently utilized at relatively low capacity in comparison with their maximum potentials. Of the three electric power supply systems, Taipower operates at the highest capacity factor, trading flexibility for lower generation cost. On the other hand, Japanese utilities and KEPCO have experienced overcapacity relative to demand.

To economize and achieve operational reliability in generation and adaptability to the load patterns, the three electricity generating systems have all opted for fossil fuel thermal units (except gas turbines that burn light oil in Taipower) to provide for the base and cycling loads, and hydroelectric units (both conventional and pumped storage) to bear medium and peak loads. Owing to inaccurate load estimates and the resulting unexpected outages, reserves are needed to meet shortages. These are met primarily from the 90–95 percent loaded units and partially from cold standbys. This has accounted for the high capacity factors for baseload units as well as the lower capacity factors for cycling and peaking units shown in Table 4.5.

Table 4.5 also shows that during the period 1955–1980 capacity factors of hydroelectric and pumped storage units have been declining continuously. It should be noted that although pumped storage was introduced in apprecia-

Table 4.4
Net Generation by Type of Load-Peaking Unit as a Percentage of Total Generation, 1965–1980

	Cycling and Peaking Unit	Medium Load Unit	Baseload Unit
Japanese utilities (nine companies)			
1965	42.4	1.6	56.0
1970	24.9	0.9	74.2
1975	21.6	7.0	71.4
1980	18.1	10.5	71.4
KEPCO			
1965	1.4	23.2	75.4
1970	4.0	13.9	82.1
1978	6.8	5.8	87.4
1980	1.4	5.3	93.3
Taipower			
1965	22.1	—	77.9
1970	15.1	—	84.9
1975	19.4	—	80.6
1980	7.4	—	92.6

Sources: Ke et al. (1983); Kim et al. (1983); Kimura et al. (1983).

— Not available.

Table 4.5
Capacity Factor by Type of Plant, 1950–1980

	1950	1955	1960	1965	1970	1975	1980
Japanese utilities							
System total	50.8	51.1	56.2	52.4	59.5	47.4	45.3
Source/type							
Oil	NA	57.1	79.8	66.8	86.0	50.1	43.7
Coal	NA	30.9	55.9	42.7	33.1	31.4	49.2
Nuclear	NA	NA	NA	NA	39.5	43.4	60.3
Hydro and pumped hydro	61.6	61.0	51.5	52.4	44.4	38.1	34.1
Natural gas	NA	NA	NA	18.2	98.0	84.7	54.1
KEPCO							
System total	18.9	31.7	50.0	45.2	39.9	45.3	43.0
Source/type							
Oil	14.1	36.0	—	21.0	35.1	45.8	42.9
Coal	21.5	15.2	52.8	52.1	53.2	56.1	61.4
Nuclear	NA	NA	NA	NA	NA	NA	63.2
Hydro and pumped hydro	17.1	47.1	46.0	37.6	42.2	30.7	18.5
Natural gas	NA	NA	NA	NA	NA	NA	NA
Industrial cogeneration	—	—	—	—	—	—	60.1
Taipower							
System total	43.1	45.5	58.4	62.1	55.4	49.3	51.4
Source/type							
Oil	—	45.1	54.8	84.1	61.6	62.9	51.2
Coal	14.3	31.0	68.7	85.0	81.2	17.9	67.3
Nuclear	NA	NA	NA	NA	NA	NA	73.6
Hydro and pumped hydro	50.2	49.7	52.6	46.9	36.0	43.8	24.3
Natural gas	NA	NA	NA	28.4	50.8	44.3	—

Sources: Ke et al. (1983); Kim et al. (1983); Kimura et al. (1983). NA Not applicable. — Not available.

Table 4.6
Largest Unit (fossil or nuclear) in the System, 1950–1982 (MW)

Year	China (Taipei)	Japan
1950	35	—
1955	40	66
1960	75	265
1965	125	350
1970	300	600
1975	500	1,000
1980	636	1,000
1982	985	1,000

Sources: Ke et al. (1983); Kim et al. (1983); Kimura et al. (1983).
— Not available.

ble quantities to the supply system in 1955 by Japanese utilities and in 1980 by KEPCO, its installed capacity has been relatively small compared with that of conventional hydroelectric plants. (In 1984 Taipower put its first pumped-storage system into operation.) The decline in capacity factor, therefore, is mostly attributable to conventional hydroelectric units.

By nature, hydropower available for electricity generation is limited. As the system expands and acquires more baseload units, it has always been advisable to position hydroelectric generation higher and higher on the load curve, resulting in greater power over shorter periods. Unfortunately, in the case of the three countries under consideration here, there have been constraints, such as limited storage owing to geographic factors and limited scale site development, which is the result of planning concepts dating back some 30 years or so. The experience of the past is instructive: During the evolution of an electricity supply system, hydroelectricity is often better suited for expanding system capacity rather than production. In addition, provision should be made to enable staged enlargement of sites.

During the course of capacity expansion, the three countries were confronted with how to make generation additions keep pace with forecasted demand, which had been rapidly rising. Aside from financial concerns, there was the problem of recruiting enough personnel to manage construction and operation. One of the solutions was to adopt large unit sizes and occasionally identical design (see Table 4.6). Simultaneously, this has permitted higher plant efficiency and better management. Table 4.7 shows the improvement in the fuel efficiency of steam power plants as more large-size units are integrated into the system. Unit size should not be so large though that it becomes incompatible with the system load, particularly when the supply system is isolated as in the cases of KEPCO and Taipower.

Table 4.7
Fuel Efficiency of Steam Power Plants, 1950–1980, by Percentage

Year	China (Taipei)	Japan	Korea	
			Oil	Coal
1955	23.4	—	—	—
1960	26.5	31.9	25.5	26.0
1965	31.4	37.1	35.1	26.7
1970	34.4	37.8	35.1	27.9
1975	35.0	38.0	36.5	31.3
1980	35.8	38.1	37.4	31.0

Sources: Ke et al. (1983); Kim et al. (1983); Kimura et al. (1983).
— Not available.

Transition in supply patterns: transmission and distribution. During the period 1950–1980 the ratio of transmission line length to generation continued to decrease rapidly in Japan, China (Taipei), and Korea (see Table 4.8). In Korea the relative magnitude fell sharply from 3.3 circuit-km per GWh in 1960 to 0.35 in 1980. Among the three countries, however, Korea maintained the highest transmission capacity on a circuit-km/GWh basis in 1980. In general the relative magnitude of the transmission capacity to generation is determined by the spatial distribution of loads and generators, the use of high-voltage transmission lines, and geographic factors. The first two are amenable to control by the utility. Since transmission and distribution losses were about the same among the three countries in 1980, Korea's greater transmission capacity relative to generation apparently can be attributed to geographic factors.

As Table 4.9 indicates, the sharp decline in losses has been attributable more to distribution than to transmission in the case of China (Taipei) and Korea.

Table 4.8
Relative Magnitude of Transmission Capacity to Generation, 1950–1980 (c-km/GWh)

Country	1950	1955	1960	1965	1970	1975	1980
Japanese utilities	1.42	1.15	0.68	0.47	0.30	0.25	0.23
KEPCO	—	—	3.25	1.96	0.83	0.35	0.35
Taipower	2.40	1.45	0.95	0.55	0.32	0.27	0.19

Sources: Ke et al. (1983); Kim et al. (1983); Kimura et al. (1983).
— Not available.

Table 4.9
Transmission and Distribution Line Losses, 1950–1980, by Percentage

	China (Taipei)			Japan			Korea		
	Trans-mission	Distri-bution	Total	Trans-mission	Distri-bution	Total	Trans-mission	Distri-bution	Total
1950	—	—	31.1	—	—	25.3	—	—	—
1955	—	—	23.9	—	—	18.4	—	—	—
1960	8.3	6.3	13.6	—	—	11.3	11.2	18.2	29.4
1965	6.1	6.0	12.1	—	—	8.5	11.4	7.9	19.2
1970	6.2	2.9	9.1	—	—	6.8	6.9	4.9	11.8
1975	4.7	2.4	7.1	3.2	5.7	6.4	6.1	5.2	11.3
1980	5.3	1.5	6.8	2.9	4.7	5.8	3.9	2.8	6.7
1983	5.1	1.7	6.8	—	—	—	3.6	2.7	6.3

Sources: Ke et al. (1983); Kim et al. (1983); Kimura et al. (1983).

— Not available.

The reasons are revealed by the sharp increase of the relative magnitude of the distribution system compared to generation as shown in Table 4.10.

As Table 4.11 indicates, all three countries worked to optimize their transmission and distribution systems. This is evident in the increase of substation capacity per unit of generation, which appeared after 1970. As a result of the sharp increases in fuel prices in 1973–1974, the three countries have made enormous efforts to raise their substation capacity, thereby decreasing transmission and distribution losses and hence fuel costs. The per unit capacity differences among them may be caused by their own geographic factors.

Table 4.12 points to the continued strong decline in the length of distribution line per unit of generation in the three countries. The most important factor for this decline appears to be the persistent increase in electricity consumption per customer. When the physical intensity of the distribution grid is standardized by customer consumption, the resulting intensity increases over time in all countries (see Table 4.10).

It is apparent that each country has made considerable efforts to extend electricity to previously isolated areas and to reinforce the existing distribution grid so that the quality of service could be improved from 1950 to 1980. More significantly, KEPCO's ambitious program of rural electrification has basically contributed to a sharp increase in the intensity of distribution facilities. This is due to the low demand that typically accompanies the first years of electrification in villages. As Figure 4.1 indicates, the peak years for Korea's rural electrification program when the annual rate of household additions exceeded 20 percent were 1965–1976. China (Taipei), on the other hand, reached a similar level of electrification (99.7 percent) from 1954 to 1973 with a relatively constant number of villages (125) becoming electrified each year.

Although data are not available about the unserved energy demand associated with outages, preliminary comparisons of the adequacy of the transmission and distribution system can be made between KEPCO and Taipower (see Table 4.13). Taipower's relatively larger number of transmission outages may be associated with high winds and typhoons. On the other hand, KEPCO's higher number of distribution outages can be attributed to the rapid rate of rural electrification.

Load characteristics. Previously the major load component in newly established electricity supply systems was residential and commercial consumption, but it was gradually outweighed by industrial consumption as industrialization and modernization flourished. During the period 1950–1980, the Japanese utilities, KEPCO, and Taipower all enjoyed high system load factors as mentioned in Chapter 3. This can be attributed to high shares of consumption in the industrial sector. Owing in part to a deterioration of system load factors, the consumption share of the industrial sector began to shrink from 1960 in Japan, from 1962 in China (Taipei), and from 1976 in Korea.

Table 4.10
Relative Magnitude of Distribution System to Generation, 1950–1980

Country	1950	1955	1960	1965	1970	1975	1980
Japanese utilities	—	3.29	3.95	4.57	5.38	5.82	5.79
KEPCO	—	—	1.42	1.39	1.82	4.44	7.43
Taipower	1.13	1.42	2.25	1.49	3.83	4.73	5.89

Sources: Ke et al. (1983); Kim et al. (1983); Kimura et al. (1983).

Note: Adjusted for consumption factors; km per capita MWh.

— Not available.

Table 4.11
Relative Magnitude of Installed Substation Capacity to Generation, 1950–1980 (MVA/GWh)

Country	1950	1955	1960	1965	1970	1975	1980
Japanese utilities	0.44	0.44	0.39	0.43	0.42	0.58	0.66
KEPCO	—	—	0.75	0.57	0.39	0.36	0.53
Taipower	0.54	0.44	0.44	0.33	0.35	0.54	0.55

Sources: Ke et al. (1983); Kim et al. (1983); Kimura et al. (1983).

— Not available.

Table 4.12
Relative Magnitude of Distribution System to Generation, 1950–1980 (km/GWh)

Country	1950	1955	1960	1965	1970	1975	1980
Japanese utilities	7.74	6.11	3.71	2.66	1.76	1.52	1.30
KEPCO	—	—	22.98	13.54	7.59	9.44	8.65
Taipower	11.69	7.96	7.00	3.12	4.34	3.30	2.53

Sources: Ke et al. (1983); Kim et al. (1983); Kimura et al. (1983).

— Not available.

Figure 3.1 in Chapter 3 indicates sharp changes in load demand throughout the day, particularly in the morning and during lunch time. Compared with only one daily peak in earlier days, this seems to have been associated exclusively with the industrial sector in which job shifts adhere to a rigid daily schedule as discussed in Chapter 3. The hourly rate of load variation can be as high as 20 percent of daily maximum, implying that in planning the mix of generation additions, it is a good idea to build in more responsiveness and adaptability.

Another issue concerning system load characteristics is the seasonal shift from winter peak to summer peak. This may have provided a certain advantage from the operational point of view. Since the summer season is usually the rainy season, and hydroelectric capacity was still maintaining 10 to 15 percent of total generation capacity composition in the late 1970s and early 1980s, hydroelectric output could be well utilized during the summer months to reduce the output required from the fossil fuel units. This advantage is now shrinking as the proportion of hydropower falls. Heavy investment in both hydropower and fossil fuels has still been needed to meet the challenge of the summer peak.

System operation. KEPCO and Taipower, as well as each of the Japanese utilities, have an automatic load dispatching center with full responsibility for system operation. Its major functions include real-time data acquisition

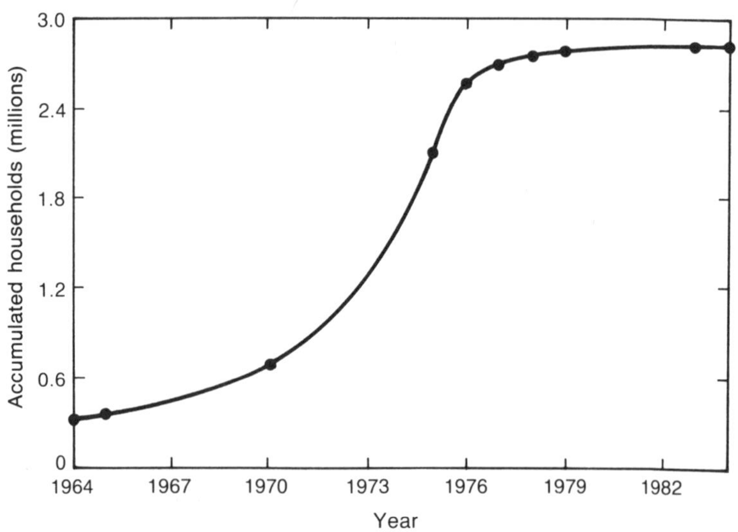

Figure 4.1. Historical trend of rural electrification in Korea. By 1976 virtually all households had been electrified.

Table 4.13
Outages of Transmission and Distribution, 1950–1980

Country	1950	1955	1960	1965	1970	1975	1980
KEPCO							
Transmission (outages/10^2 c-km)	—	—	7.64	6.01	3.46	2.03	0.85
Substation (outages/10^2 MVA)	—	—	12.90	12.90	4.99	1.97	0.30
Distribution (outages/10^2 c-km)	—	—	47.20	66.70	27.90	7.90	3.20
Taipower							
Transmission (outages/10^2 c-km)	1.00	1.61	1.31	6.16	7.40	8.50	5.03
Substation (outages/10^2 MVA)	—	10.40	3.64	0.80	0.46	0.33	0.17
Distribution (outages/10^2 c-km)	—	33.20	17.20	4.74	3.20	2.68	2.74

Sources: Ke et al. (1983); Kim et al. (1983); Kimura et al. (1983).

— Not available.

and processing, automatic generating control, system monitoring, voltage and reactive power control, and interchange of power between regions within its service area and between neighboring systems (in Japan) to balance demand with supply and operate at the lowest possible cost. Both KEPCO and Taipower have computerized dispatching; Leeds and Northrop analog frequency control equipment was acquired and made operational for both utilities in the late 1970s. One drawback that has hindered smooth and effective system operation is the mismatch between excess concentration of power sources in one region and load demand in that region. Another is the inflexibility of nuclear output, which has caused hardship during off-peak operation.

Maintenance and inspection of generation facilities are carried out by the maintenance center with the assistance of the respective plant staff. The management of transmission lines maintenance is designated among the line superintendants of the electrical system department, as in the case of Taipower. KEPCO contracts both functions to Korea Heavy Industries and Construction Company, Ltd. Table 4.14 shows the scheduling of maintenance and inspection as well as the maintenance period duration for the generating facilities set by Taipower and KEPCO. It has been a general practice every three to four years to take advantage of the dry season to carry out major maintenance and inspection of hydroelectric units.

Good management of maintenance and inspection greatly improves system reliability. The advanced technology employed in system operation and well-established management in maintenance and inspection have brought about a sharp decrease in outages of transmission, substations, and distribution for both KEPCO and Taipower during the period 1950–1980, as Table 4.13 shows. And, as Table 4.15 indicates, there is evidence that the outage rates and availability of the generation facilities have been kept at a better-than-acceptable level by both KEPCO and Taipower.

EXPANSION PLANNING

The second oil shock in 1979 brought basic changes to the electric supply industries in China (Taipei), Japan, and Korea, reducing future prospects for economic growth and income elasticity of electricity consumption. In addition to lower demand growth, the unbalanced growth of peak and baseload demands, which became increasingly apparent after the second oil shock, has been accentuated.

The second oil shock and the international economic recession that followed were not anticipated, and it is evident that the three countries are now faced wth excess generating capacities. Worse yet, the electric utilities appear to be encountering a peculiar situation of excess capacities with shortfalls of peaking services. Even though they have revised their electric power expansion programs in view of the future prospect of lower demand growth, the revisions actually would have no effect on their installed capacities before the late 1980s. Most capacity additions through the late 1980s were already under construction in early and mid decade because fossil-thermal and nuclear baseload generation facilities have long lead times for investment. Accordingly, substantial excess reserve capacity is likely to persist in all three systems all through the 1980s (see Table 4.16). In addition, the heavy emphasis on coal and nuclear baseload plants without consideration of other types

Table 4.14
Scheduled Maintenance Period Duration for Generating Plants

Type of Plant	Days/Year
KEPCO	
Units rated at 200 MW and higher	28–55
Units rated at 100–190 MW	21–45
Units rated below 100 MW	21–35
Taipower	
Units rated at 400 MW and higher	40–52
Units rated at 200–400 MW	30–48
Units rated at 100–200 MW	30–42
Units rated below 100 MW	25–42

Sources: Ke et al. (1983); Kim et al. (1983).

Table 4.15
Outage and Availability of Generating Facilities by Percentage

Item	1975	1980	1981
KEPCO			
Coal-fired			
forced outage	9.8	4.3	4.8
scheduled outage	10.8	13.8	15.3
availability	80.4	82.5	80.6
Oil-fired			
forced outage	6.0	7.2	6.0
scheduled outage	18.8	18.0	18.4
availability	76.4	76.1	76.7
Nuclear			
forced outage	NA	4.1	4.0
scheduled outage	NA	17.3	27.7
availability	NA	79.4	69.4
Combined coal, oil, and nuclear			
forced outage	18.2	6.4	5.7
scheduled outage	18.2	17.5	19.0
availability	76.4	77.2	76.4
Taipower			
Coal-fired			
forced outage	6.4	12.7	23.0
scheduled outage	6.7	5.9	7.0
availability	87.3	82.1	71.6
Oil-fired			
forced outage	3.85	5.5	7.55
scheduled outage	9.1	9.4	11.1
availability	87.4	85.6	82.2
Nuclear			
forced outage	NA	11.9	10.2
scheduled outage	NA	12.6	5.3
availability	NA	77.0	85.0
Combined coal, oil, and nuclear			
forced outage	4.7	8.2	10.5
scheduled outage	8.3	9.6	9.2
availability	87.4	83.0	81.3

Sources: Ke et al. (1983); Kim et al. (1983).

NA Not applicable.

Table 4.16
System Gross Reserve Rate by Percentage

Year	China (Taipei)	Japan	Korea
1980	30.9	45.2	75.4
1985	87.8	—	74.9
1990	54.5	33.4	44.6
1995	36.2	—	44.9
2000	30.4	32.2	38.5

Sources: Ke et al. (1983); Kim et al. (1983); Kimura et al. (1983).
— Not available.

of technology has caused a mismatch between load characteristics and generation mix.

The Need for Reassessment in the 1980s

Under conditions of lower economic growth, radical changes in industrial structure, built-in excess baseload capacity, and shortages of peaking capacity, the three utility systems will have to decide how to develop an optimal electric power expansion program that takes load management into account. The environment for power planning has changed considerably. During the 1970s, the electric utilities encountered increasing uncertainty in the variables related to utility management, and it was difficult for management to adjust rapidly to the changes.

In addition to the inherent uncertainty of demand projections discussed earlier, fuel costs, power plant capital costs, interest rates, and plant performance have all suffered unexpected and wild fluctuations. What was assumed to be a rather straightforward optimization problem has become a problem fraught with uncertainties that require strategic planning to maximize flexibility and adaptation to unknown future changes. The softening of the international oil market and the resulting oil price decreases in the 1980s present an even more complicated and difficult problem, as further decreases in oil prices, relative to coal prices, could erode the cost advantage of coal-fired plants. The challenges are expected to grow more acute during the remaining years of the 1980s.

The utilities are now reconsidering their capacity expansion plans. The key issue is to determine the most efficient supply structure, including transmission, that matches the demand pattern. Assuredly, the three countries of Northeast Asia need to expand their generating capacities during the rest of the century to meet expected demands for electricity. As discussed in Chapter 2, in view of the prospect of lower-than-expected total electricity demand

and the resulting drastic changes in demand pattern, the three utility systems substantially revised their electric power expansion programs in the 1980s. These changes affected the number, size, and type of planned power plants.

The construction of nuclear power plants is still given top priority because their generation cost is considered to be lowest. But even nuclear units are delayed in the new plans by six months to one year, and coal-fired and liquified natural gas (LNG) fired thermal units have been postponed by two or more years. Table 4.17 shows the development of expected generating capacity by source; Table 4.18 shows the projected evolution of electric power production by source; and Table 4.19 gives the future annual fuel requirements by source.

According to Table 4.17, between 1980 and 2000, total Japanese generating capacity is expected to increase from 129 GW to about 300 GW (an average annual growth rate of 4.3 percent). Total Korean generating capacity is expected to increase from 9.5 GW to 49 GW (an average annual growth rate of 8.5 percent), and in between is Taipower's total generating capacity, which is projected to grow from 9 GW to 24.5 GW (an average annual rate of 5.1 percent). Energy policy in China (Taipei), Japan, and Korea stresses nuclear

Table 4.17
Future Planned Generating Capacity by Energy Source, 1980–2000

	System Total (MW)	Source/Type (%)				
		Oil	Coal	Nuclear	Hydro[a]	Natural Gas
Japanese utilities						
1980[b]	129,360	46.5	4.1	12.0	22.3	15.2
1981[b]	135,502	46.6	4.4	11.9	22.6	14.5
1990	209,000	24.1	11.0	22.0	22.4	20.6
2000	300,000	15.0	17.0	30.0	22.0	16.0
KEPCO						
1980[b]	9,574	72.0	7.8	6.1	12.1	0
1981[b]	9,835	74.2	7.6	6.0	12.2	0
1991	23,730	16.7	19.1	39.7	13.8	10.7
2000	49,000	12.2	22.9	44.5	10.2	10.2
Taipower						
1980[b]	9,056	59.8	10.8	14.1	15.3	0
1981[b]	10,159	54.5	9.6	22.2	13.7	0
1985	15,790	34.1	19.1	32.6	14.2	0
1990	17,000	30.4	22.3	30.2	17.1	0
2000	24,500	18.9	19.0	41.4	20.7	0

Sources: Ke et al. (1983); Kim et al. (1983); Kimura et al. (1983).

a. Includes pumped hydro and geothermal power.

b. Actual.

power as one of the most effective means of reducing dependence on oil imports. In principle plans in all three countries call for no more oil-fired units to be constructed. Whereas there is an obvious need to promote the full use of all economically available hydropower and to look seriously at the future role to be played by new forms of energy generation in these countries, the nuclear option is thought to be the most viable way to meet the growing need for reliable and economical power by the end of the century. Accordingly, future generating facilities plans call for ambitious nuclear programs, the proportion of nuclear units in total installed generating capacity for the year 2000 ranges from 30 percent in Japan to 41 percent in China (Taipei), and to 45 percent in Korea (see Table 4.17).

As discussed in Chapter 5, high-quality construction and operating personnel and procedures are essential to ensure satisfactory performance of nuclear power plants, which is inducing KEPCO and Taipower to consider further scaling down their nuclear programs. Switching to less problematic generation technologies such as coal might be advisable, if some difficulties exist in creating independent capacity in nuclear regulation, personnel training, and equipment supply and maintenance.

Table 4.18
Future Planned Generation by Energy Source, 1980–2000

	System Total (GWh)	Source/Type (%)				
		Oil	Coal	Nuclear	Hydro[a]	Natural Gas
Japanese utilities						
1980[b]	514,100	44.8	4.4	16.0	16.7	15.0
1981[b]	523,144	47.1	5.5	16.7	16.2	14.5
1990	840,000	19.0	12.4	30.1	14.8	21.3
2000	1,200,000	11.0	16.0	43.0	15.0	15.0
KEPCO						
1980[b]	36,047	71.9	11.2	9.0	5.2	0
1981[b]	40,207	79.8	6.3	7.2	6.7	0
1991	145,509	11.3	19.5	50.8	3.5	14.9
2000	210,784	10.8	21.8	55.4	0.9	11.1
Taipower						
1980[b]	40,813	59.6	14.2	19.1	7.1	0
1981[b]	39,644	50.0	11.1	27.0	11.9	0
1985	48,765	9.1	27.9	52.5	10.5	0
1990	66,709	15.5	29.1	43.9	11.5	0
2000	114,671	11.3	28.5	48.5	11.7	0

Sources: Ke et al. (1983); Kim et al. (1983); Kimura et al. (1983).

a. Includes pumped hydro and geothermal power.

b. Actual.

Radically different mixes of generation are expected in the year 2000 than currently exist. By that year, according to Table 4.18, nuclear reactors will generate 49 percent of all electricity consumed in China (Taipei), 43 percent in Japan, and 55 percent in Korea. In order not to become overly dependent on any one power source, Japanese utilities and KEPCO are also rapidly turning to coal and natural gas. In Taipower's planned generation mix, nuclear and coal capacity is to be split more evenly by the year 2000 with no gas units. As a result, between 1980 and 2000 the proportion of total electricity supply generated by oil-fired units will sharply decline from 60 percent to 11 percent in Taipower, from 45 percent to 11 percent in the Japanese utilities, and from 72 percent to 11 percent in KEPCO (see Figure 4.2). This represents a conscious government effort in these countries to phase out oil as a fuel for electricity generation, thereby pulling overall oil dependency downward.

To some extent, the continuing excess capacities can be economically justified during the 1980s, as long as baseload generation is cheaper than the fuel costs of existing oil-fired plants. Projections of continuing low oil prices could encourage a delay in the closing of oil-fired plants. Substantial construction of new oil-fired plants is unlikely. The utilization of existing oil-fired plants should not drop below a capacity factor of 30 to 40 percent, however, or there will be a substantial burden on the economy. In particular, in the case of Tai-

Table 4.19
Future Annual Fuel Requirements, 1980–2000

	Oil (10^3 kl)	Coal (10^3 ton)	LNG (10^3 ton)	U_3O_8 (tons) (3% enrichment)
1980				
Japanese utilities	53,240	9,780	13,440	410
KEPCO	7,090	1,645	0	14
Taipower	6,039	2,422	0	39
Total	66,369	13,847	13,440	463
1990				
Japanese utilities	41,200	42,000	31,500	1,265
KEPCO (1991)	2,709	8,088	2,700	294
Taipower	2,180	8,091	0	177
Total	46,089	58,179	34,200	1,736
2000				
Japanese utilities	34,000	77,000	31,700	2,600
KEPCO	2,476	17,160	3,524	660
Taipower	3,359	13,267	0	360
Total	39,835	107,427	35,224	3,620

Sources: Ke et al. (1983); Kim et al. (1983); Kimura et al. (1983).

Note: 1.0 kl = 6.3 barrels. All tons are metric.

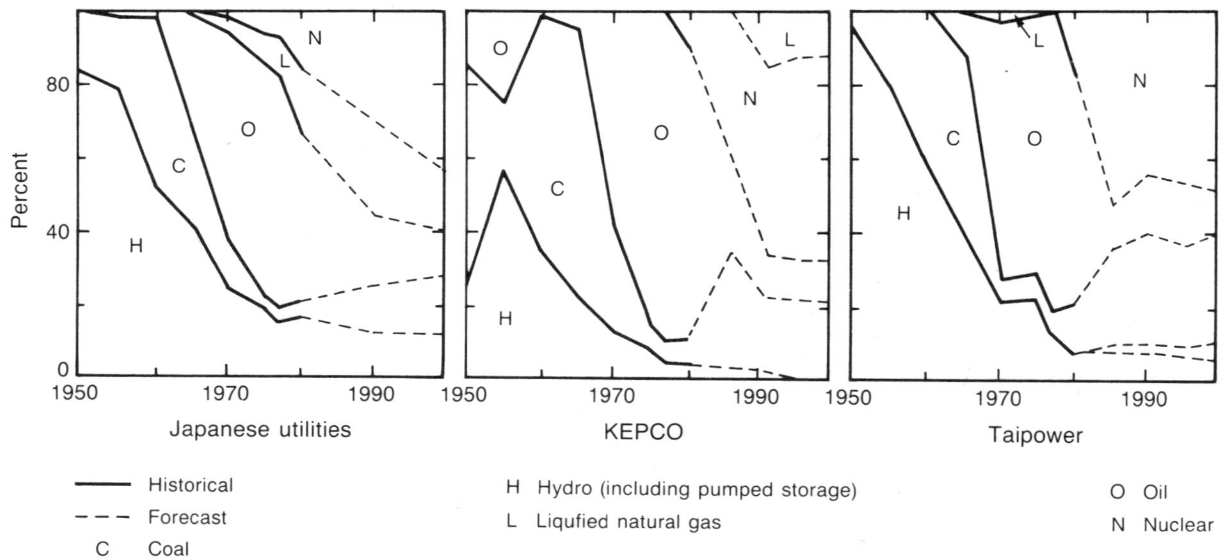

Figure 4.2. Historical and forecasted electricity generation by fuel in Northeast Asia.

power the capacity factor of oil-fired plants has been much lower than 12 percent during the 1980s, which means that excessive substitution of new coal and nuclear baseload generation for existing oil-fired plants has occurred if the load growth rate does not turn around. Faced with excess baseload capacity, one of the critical issues facing the electric utilities in the three countries is how to improve the load duration curve—that is, how to create new baseload demand—and to shift part of peak demand to off-peak demand through special sales promotions, time-of-use rates, and other load management techniques (see Chapter 3).

Considerable quantities of coal, LNG, and uranium will be required for power generation by the three countries. As Table 4.19 shows, the annual requirement for steam coal in these countries is forecast to increase sharply by approximately eight times, from 14 million metric tons in 1980 to 107 million metric tons in the year 2000. During the same period, the estimated annual LNG requirement will rise from 13 million to 35 million metric tons, and the annual requirement for uranium will grow by about eight times, from 463 metric tons in 1980 to 3,620 metric tons in the year 2000. Under these conditions, oil requirements by the three countries will sharply decline from 66 million kl to 40 million kl (415 to 250 million barrels/year) over the same period (see Table 4.19). In addition to oil, almost all steam coal, LNG, and uranium required by China (Taipei), Japan, and Korea will have to be imported from nondomestic sources. In the early 1980s the importation of fuels constituted more than US$10 billion of business per year for the fuels alone.

Technology Options: Prospects and Costs

Considering the reliability of fuel supply, as well as cost effectiveness, the task of the electric utilities is to determine the optimal combination of technologies capable of adapting to a given load shape. Baseload generation accounts for a major portion of total generation cost, and the minimization of baseload generation cost is thus one of the utility's critical management issues.

Since the three countries of Northeast Asia are all deficient in domestic energy resources and heavily dependent on imported oil, the choices for baseload power generation in the future are basically among nuclear and coal-burning technology. In these countries nuclear technology is viewed as a more economical baseload supplier than coal, a semidomestic source of energy (once an independent nuclear fuel cycle is established), and a relatively reliable fuel source. Moreover, the establishment of nuclear industry is closely related to a nation's industrial development strategy. These positive aspects of nuclear technology must be weighed against safety and waste-disposal problems as well as growing costs. High technical competence is also needed for ensuring safe operations of nuclear facilities (see Chapter 5).

The Japanese utilities, KEPCO, and Taipower have provided an estimate of comparative costs for oil-fired, LNG-fired, coal-fired, and nuclear generation for the period in question (see Table 4.20). The international cost comparisons show that nuclear generation could have a cost advantage of 10–20 percent over coal, 35–40 percent over LNG, and 35–40 percent over oil. With these calculations, nuclear technology is indeed the cheapest source for power generation.

Coal first realized a cost advantage over oil during the period 1979–1980. With oil prices more than doubling, the fuel cost weight of thermal units in total generation cost has sharply increased. The increase in oil prices has

Table 4.20
Cost Comparisons of Alternative Baseload Generation

	Oil	LNG	Coal	Nuclear
Japanese utilities[a]				
Construction cost (US$/kW)	522	658	840	1,112
Generation cost (US cents/kWh)	8.85	7.95	6.58	5.22
fuel cost (US cents/kWh)	7.08	5.96	3.62	1.31
	(0.80)[b]	(0.75)	(0.55)	(0.25)
fixed cost (US cents/kWh)	1.77	1.99	2.96	3.92
	(0.20)	(0.25)	(0.45)	(0.75)
KEPCO[c]				
Construction cost (US$/kW)	637	502	877	1,506
Generation cost (US cents/kWh)	8.05	8.48	5.67	4.98
fuel cost (US cents/kWh)	5.98	6.98	2.70	0.77
	(0.74)	(0.82)	(0.48)	(0.15)
fixed cost (US cents/kWh)	2.07	1.50	2.97	4.21
	(0.26)	(0.18)	(0.52)	(0.85)
Taipower[d]				
Construction cost (US$/kW)	725	NA	882	1,238
Generation cost (US cents/kWh)	6.41	NA	5.19	4.03
fuel cost (US cents/kWh)	4.79	NA	3.09	0.79
	(0.75)	NA	(0.60)	(0.20)
fixed cost (US cents/kWh)	1.62	NA	2.10	3.24
	(0.25)	NA	(0.40)	(0.80)

Sources: Ke et al. (1983); Kim et al. (1983); Kimura et al. (1983).
a. 1981 prices.
b. Numbers in parentheses indicate shares in the total.
c. 1982 prices.
d. 1983 prices (after reduction of oil price).
NA Not applicable.

dragged up the prices of LNG and coal during the period 1979–1980. Consequently, in the three countries, as Table 4.20 shows, the fuel cost weight for nuclear power is the lowest, followed by coal, oil, and, finally, LNG. It is obvious that nuclear technology is attractive to these countries since most of their primary energy sources are imported from abroad.

Although detailed cost studies indicate that in China (Taipei), Japan, and Korea, nuclear power shows an advantage over direct coal-fired units, they are both competitively viable options, and the future cost uncertainties do not give one a significant advantage over the other. Considerable uncertainties prevail with respect to the relative prices of oil, LNG, and coal, and also the capital cost of nuclear power. Accordingly, the three countries consider coal-burning technology as an economical baseload supplier next to nuclear and a more secure fuel compared with oil.

Like nuclear technology, however, coal-burning technology raises serious environmental issues. As Figure 4.3 shows, Japan has made monumental strides in controlling such impacts, with the average sulfur content of fossil fuels burned by the Tokyo Electric Power Co. (TEPCO) dropping dramatically between 1965 and 1977. Since then, however, further progress has been difficult to achieve. This is reflected in the air pollution statistics for the country as well, which have leveled off after a period of improvement (see Figure 4.4). Looking to the future, new technologies to utilize coal, such as fluidized bed and synthetic fuel, would be attractive in Japan because the enormous amount of coal located in the Asia-Pacific region can be used only when significant improvements over present systems are economically available for pollution control.

The softening of the international oil market and the oil price decreases of the 1980s further complicate the problem. The electric utilities in the three countries are now bewildered about their basic directions in power development planning. As Table 4.20 indicates, even though the total cost of coal-fired generation is substantially lower than that of oil-fired generation in the Japanese utilities and KEPCO, this cost advantage was noticeably reduced in Taipower in 1983 by the oil price decrease. Further price decreases have brought oil and coal quite close together. If the soft condition of oil supply and demand continues, and coal prices show no sharp decreases in the coming years, oil power generation may again turn out to be more economical than coal technology. Furthermore, the future excess surplus of residual fuels and the resulting depression of their prices could tilt the advantage further in favor of oil technology (see Chapter 7).

Existing oil-fired baseload generation has in some cases already proved more economical than the new coal-fired baseload generation since the capital costs of existing oil-fired power plants can be regarded as "sunk." On the other hand, it would be difficult to argue for major new expansions of oil-fired capacity because of the history of oil supply price and stability. Thus in the presence of uncertainties regarding the future course of relative fuel costs, the basic policy of electric power development should be directed to the diversification of fossil fuel sources among oil, LNG, and coal with reasonable reduction of present oil dependency.

In an attempt to diversify power sources further, the use of LNG will be greatly increased in both Japan and Korea. LNG, a clean-burning fuel, is easily stored around metropolitan areas, thereby allowing for effective transmission. A relatively secure supply can be guaranteed from long-term contracts that usually involve the whole system of production, shipping, receiving, and distributing facilities. Accordingly, LNG-burning technology is highly capital-intensive and inflexible in supply due to the so-called take-or-pay clause that is prevalent in the industry.

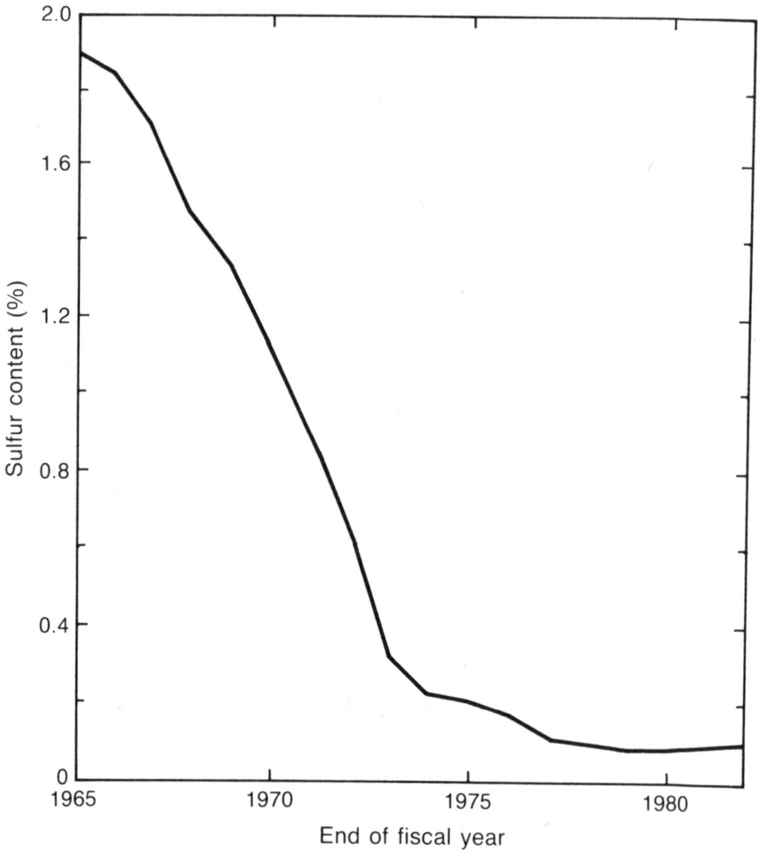

Figure 4.3. Historical decline in average sulfur content of Japanese power plant fuels.

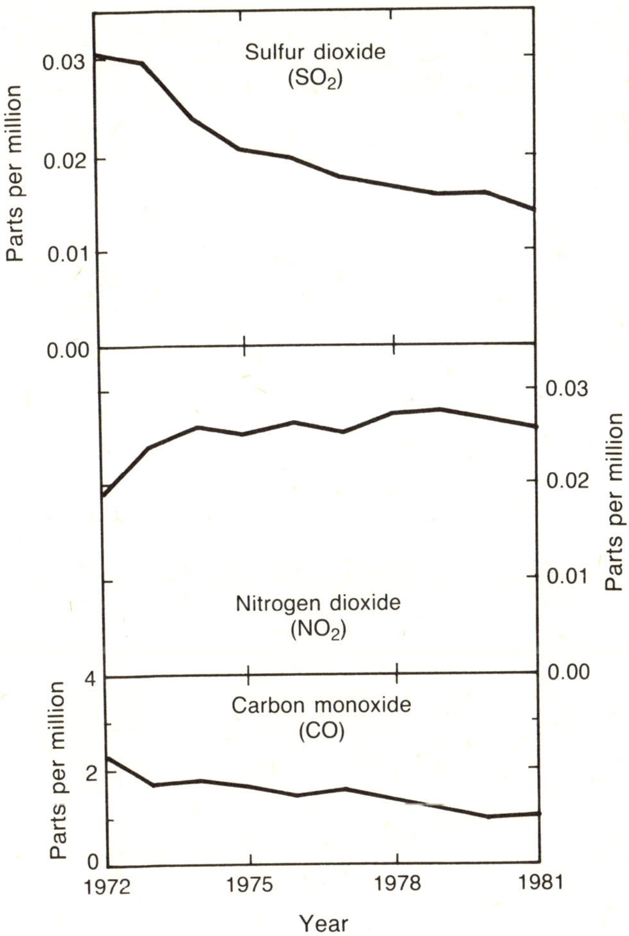

Figure 4.4. Historical change in annual average air pollution concentrations in Japan. SO_2 concentrations decreased more than CO, and NO_2 actually increased during the 1970s. SO_2 is mostly formed by stationary combustion such as power plants, whereas the other two are also formed in more difficult-to-control mobile sources such as automobiles.

In China (Taipei) the introduction of gas-fired units into the system is still under review because it requires costly infrastructure in maritime transportation facilities, liquefaction plants, pipeline networks, and pumping stations. Natural hazards such as high winds and typhoons are also constraining because they limit sites for LNG ports. An LNG project in Korea has been initiated in an effort to alleviate residential fuel problems in metropolitan areas (city gas) and to diversify fuel sources of power generation.

In the future, Japan hopes to achieve higher efficiency in the use of LNG, partially through the utilization of thermal discharges and recovery of the cold energy of liquefaction and partially through the introduction of new combustion technologies such as combined cycles. Indeed, in 1987 TEPCO started operation at one of the world's most advanced and, what will become at 2,000 MW, largest combined-cycle power plants, with an overall thermal efficiency of 44 percent and ability to follow load (International Power Generation, 1987a).

Liquified petroleum gas (LPG) is readily available in international markets, but its supply and demand is unstable, resulting in frequent price fluctuations. Consequently, Japan plans to utilize LPG as a supplementary fuel to LNG for siting and environmental considerations.

An exciting new prospect for urban power generation is being tested in Tokyo: gas-powered fuel cells (Marshall, 1984). These offer the possibility of achieving clean and efficient power production with the possibility of utilizing the waste heat. In addition to the attractiveness of siting flexibility, the fuel cell offers unit-size and throttling flexibility. It can achieve 40 percent efficiency of power generation at low power and at small size (conventional thermal plants only approach this level at full power in large units). Finally, unlike most other power systems, the construction time for fuel cells can be quite short, maximizing utility flexibility in capacity planning. In an age of many uncertainties, the fuel cell offers attractive possibilities. Its uncertainties are in the useful life of the product and capital costs, factors that are now under study. Should the fuel cell prove as effective as its backers hope, gas could be much more in demand as a power plant fuel.

Gas is also a particularly good fuel for use in cogeneration plants, which are often designed to have multifuel capability. It can be expected that industrial and commercial cogeneration will grow rapidly in Northeast Asia as they have in the United States where recent estimates indicate that as much as 20 percent of total electricity may be supplied in this manner by the year 2000 (Clark, 1986).

Finally, hydropower in China (Taipei), Japan, and Korea will be developed mainly for pumped storage in order to utilize off-peak generation of the baseload equipment to achieve better fuel economy, whereas other renewable resources, such as solar, wind, ocean energy, biomass, and geothermal, appear not to be as economical as large-scale commercial operations in the near future. A possible exception is tidal and tidal current power, which are under serious consideration in Korea (International Power Generation, 1987b).

Although the future course of relative fuel prices is a critical element in evaluating the probable life-cycle generation costs of the alternative technologies, capital costs have as significant an impact on total generation costs as the choice of technology. Furthermore, capital costs have increased sharply in the past decade, though to a lesser degree than fuel prices. Large uncertainties exist with respect to future capital cost movements. In this respect, nuclear technology has some disadvantages, as the capital cost weight of nuclear generation in total generation costs is the highest. On the other hand, the increase in fixed costs is mainly caused by the increase in plant construction costs and the decrease in the plant capacity factor.

Relatively little systematic information is available on the construction cost increases of power facilities in the three countries. It is known, however, that construction costs of nuclear plants have increased in real terms by 5 to 10 percent per year over the past decade in most industrialized countries. The increase for coal plants is also substantial, though to a lesser degree than nuclear plants, as Figure 4.5 indicates. In Table 4.20, fixed costs of nuclear plants in China (Taipei), Japan, and Korea turn out to be 32 to 54 percent higher than those of coal-fired plants. Whereas investment in pollution controls accounts for most of the construction cost increases of coal-fired plants, lengthening of construction periods has been a major cause ascribed to the sharp increases in construction costs of nuclear plants. In the case of the three countries, construction periods are six to eight years for nuclear plants and four to five years for coal-fired plants. Accordingly, the effect of interest costs on capital costs during construction is greater for nuclear plants than for coal-fired plants. Since higher safety and increased quality control have increased the construction periods of nuclear plants, utility management should reduce construction costs by promoting the standardization of plants. This can be accomplished while maintaining strict standards of safety and reliability.

The capacity factor also has a significant impact on the unit charge for capital costs and hence the choice of technology. There are substantial uncertainties about performance of nuclear plants as a result of less than satisfactory operating experience. Whereas Japanese nuclear plants accomplished a satisfactory capacity factor of 60 percent in 1979 and 1980, and 61 percent in 1981 on the average, in Korea the first nuclear power plant achieved capacity factors of 46, 61, 67, and 56 percent in the first four years of operation (1978–1981). During the same period in China (Taipei), nuclear operation achieved capacity factors of 44, 57, 74, and 81 percent (see Chapter 5). More recently, however, capacity factors have risen.

High-quality construction and operation personnel and procedures are essential to ensure satisfactory performance of nuclear plants. If real construction costs of nuclear plants continue to increase significantly, along with a recurrence of the recent low capacity factors, constant coal prices in real terms could substantially reduce the cost advantage of nuclear plants over coal-fired baseload plants. Indeed, in the mid-1980s TEPCO generation costs for coal and nuclear electricity show signs of converging again.

In summary, the choice of technology and the consequent fuel-mix constraints imposed by the three countries are similar. They are based on the reliability of fuel supply at the lowest possible cost, the impact of the chosen technology on the nation's industrialization, and environmental and siting constraints. From a cost-effectiveness viewpoint in the three countries, around 20 percent of future demand will be supplied by hydro systems (including conventional and pumped storage); nuclear generation will be developed as

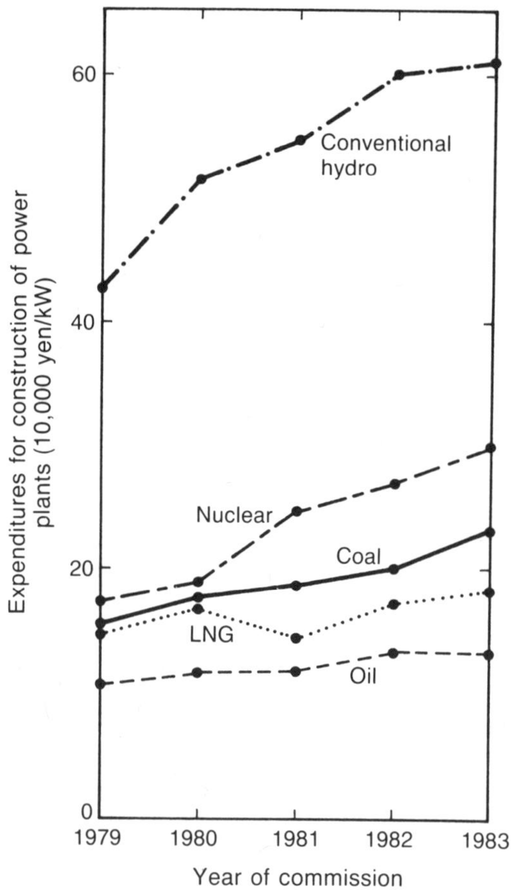

Figure 4.5. Historical trend of power plant construction costs in Japan.

much as possible to meet about 40 to 50 percent of demand; and around 30 to 40 percent will be met by fossil-thermal plants. For now, fossil-thermal plants may be part of baseload suppliers, but later they will be converted to take the intermediate load. Part of the existing fossil-thermal plants will be used by cycling services, and newly built plants, including combined cycles, will be utilized for this purpose.

In addition, fuel diversification will be pursued to cope with any potential repeat of the energy crisis. Not only nuclear technology but also LNG- and coal-fired technologies will be actively promoted. In every case, the flexibility of operation must be fully considered because naturally flexible generation, such as hydro and oil-fired plants, will account for smaller shares.

PLANNING METHODS

In an effort to achieve the most efficient possible mix of generating capacity, at present both Japanese utilities and Taipower use a probabilistic simulation model for evaluating the performance of prescribed capacity expansion plans (see Table 4.21). Considering the substantial uncertainties in demand projections, the future course of relative fuel prices, power plant capital costs, and performance of new technologies, the application of complex optimization methods may not be cost effective. This is particularly the case where the choice of plant fuels is partly predetermined politically to secure a reliable fuel supply. In light of the growing importance of fuel and operating costs in total generation costs, moreover, a probabilistic simulation model can provide more accurate configurations of fuel and operating costs.

KEPCO uses a probabilistic simulation model in conjunction with a lowest-possible-cost power system expansion model (WASP). The WASP computer code could be applied until recently because almost all of KEPCO's generating capacity was based on thermal power.

In the future, considering the unbalanced generation mix and load characteristics in the three countries, a more detailed operational representation is desirable to help the electric utilities analyze the potential for conversion of existing fossil-thermal units in ways that minimize total costs and at the same time do not cause any operational difficulty. In addition to integration of capacity expansion planning with system operation, integration is desirable with financial planning so that much more detailed information on short-term financial requirements can be available in the future.

To determine the optimal mix of generating capacity, KEPCO and Taipower use the present worth of revenue requirements as the key economic measure of efficiency, although foreign fund requirements are also evaluated. In both, systems reliability is treated as a constraint to assure the adequacy, quality, and economy of electric power supply. Both KEPCO and Taipower are presently reviewing the social costs of supply-demand mismatch for electricity.

Table 4.21
Capacity Expansion Planning Methods in Northeast Asia

	Japanese Utilities	KEPCO	Taipower
Planning method	Probabilistic simulation Prescribed path	Dynamic programming Probabilistic simulation	Prespecified path Probabilistic simulation
Performance evaluation	0.3 day/month	Present worth of revenue and foreign fund requirements Reliability constraints (0.7 day/year	Present worth of revenue and foreign fund requirements Reliability constraints (3 days/year)
Technology options			
Nuclear technology	More economical for baseload Semidomestic source More reliable fuel Market penetration of fast breeder (2010 or 2020) Safety and waste disposal problems	More economical for baseload Semidomestic source More reliable fuel Market penetration of fast breeder (2010 or 2020) Safety and waste disposal problems Skilled manpower problem	More economical for baseload Semidomestic source More reliable fuel Market penetration of fast breeder (2010 or 2020) Safety and waste disposal problems Skilled manpower problem
Coal technology	Economical for baseload More secure fuel New technology (synthetic fuel) Environmental constraints	Economical for baseload More secure fuel Environmental constraints	Economical for baseload More secure fuel Environmental constraints

LNG-burning technology	Clean energy Capital intensive Easy siting and effective transmission LPG as supplementary to LNG	Clean energy Capital intensive Easy siting and effective transmission Alleviating residential fuel problems	Clean energy Capital intensive Easy siting and effective transmission Economic infeasibility and natural hazards
Key factors to be considered	Appropriate reserve margin Optimizing generation mix Economic evaluation of each project using cost/value ratio Around 20% of increases in demand will develop for peak load, mainly supplied by pumped storage Nuclear power will be developed as much as possible Diversification of fuels Flexibility of operation	Economic cost/discount rate Reliability of fuel supply at a reasonable price Capital intensiveness and ease of financing System reliability and technological uncertainty Public acceptance and environmental concerns	Economic cost Diversification of fuel forms and sources Capital intensiveness and ease of financing System reliability and technological uncertainty Regional balance of demand and supply Standardization of unit type and size Optimal mix of generation type
Treatment of uncertainty	Annual revision of plans Sensitivity analysis Contingency planning	Annual revision of plans Sensitivity analysis Contingency planning	Annual revision of plans Sensitivity analysis Contingency planning

Sources: Ke at al. (1983); Kim et al. (1983); Kimura et al. (1983).

According to a preliminary study by a foreign research institute, the social cost of the electricity deficit in Korea amounts to somewhere between six and eight U.S. cents per kilowatt-hour. It is difficult to evaluate the social cost of the deficit resulting from the frequency of outages, the loss of jobs, and lower economic growth.

As discussed in Chapter 3, KEPCO and Taipower use a loss of load probability (LOLP) of 0.7 days per year and 3 days per year, respectively, whereas Japanese utilities use an LOLP constraint of 0.3 days per month (see Table 4.21). Taipower's system generation mix suggests a 55 to 60 percent baseload, 30 to 35 percent intermediate, and 20 percent peak, allowing 20 to 25 percent as a reserve margin. Whereas Japanese utilities use peak daily values, KEPCO and Taipower use the actual load duration curve for computation of LOLP. Since some of the causes of generation unreliability are seasonal, such as droughts, it would seem more appropriate to set a monthly reliability criterion than a yearly one.

The frequency of reliability failures in the transmission and distribution systems should be considered in setting reliability constraints for the generation system. If not appropriately considered, excess generation capacity might result in no direct benefit to the customer. Rather, given the different value of unserved energy supply to different users, management techniques aimed at improving service reliability at the distribution system level tend to provide a better payoff. In addition to the economic costs and system reliability or reserve margins mentioned above, other key factors in expansion planning are shown in Table 4.21. The main factors are similar in all three countries.

Since the oil crises in 1973 and 1979, the oil price jump has enabled alternative energy sources to be developed, and every effort has been made to expedite nonpetroleum power development programs. The three countries also intend to invest in coal and uranium exploration and development on a joint venture basis in some promising areas overseas to ensure stable supply and price.

The regional power balance is one of the basic considerations in power development planning. Regional balance of supply and demand can reduce transmission costs and line losses, as well as improve system reliability. Since imported coal will comprise an increasing percentage of the total imported energy in the future, all the relevant facilities and equipment, such as coal-unloading terminals and transportation means and equipment, will necessitate a large outlay of capital investment. In order to utilize those facilities fully, new coal-fired plants should be planned with a degree of unit concentration rather than being widely dispersed. This implies that practical application of the regional power balance should not be rigid but should be given some flexibility in view of the overall economy.

In order to promote technology transfer and foster domestic manufacturing capability of heavy electrical machinery in China (Taipei) and Korea, the standardization of unit and size will need to be vigorously pursued. In addition, this should help shorten the time required for unit overhauling and reduce the component stock and inventory.

In KEPCO, Taipower, and TEPCO, peaking load accounts for 10 to 20 percent of the daily peak load, intermediate load occupies 20 to 35 percent, and baseload takes up 45 to 70 percent of the system peak (see Chapter 3). As far as daily load fluctuation is concerned, the hourly rate of load change may be as high as 20 percent of the system peak in Japan, and progressively less in China (Taipei) and Korea. Maximum and minimum load ranges are 31 percent, 44 percent, and 56 percent of the daily peaks, respectively, in Korea, China (Taipei), and Japan. It is predictable that such phenomena will exist for quite a long period. In view of the anticipated load growth, such load changes and variation range will be rendered ever more striking; the selection of unit type and size and the optimal mix of units for flexibility of operation will, therefore, become particularly important to the growing power systems.

The Taipower system is not connected with neighboring countries. From the viewpoint of system stability, therefore, Taipower assumes that the maximum size of the unit should not be larger than 7 percent of its peak load. Interestingly, in Israel power supply is likewise isolated from its neighbors, but Israeli utilities operate with units as large as 25 percent of off-peak demand through careful switching, load shedding, and emergency backup on site. Considering the relatively high power density in China (Taipei), Japan, and Korea, it seems reasonable that the three should adopt the basic approach of locating the units as close to load center as possible, thereby reducing transmission and distribution losses and enhancing system reliability.

Since the mid-1970s, environmental impacts and scenery protection have become sensitive issues imposed on new power generation establishments, particularly nuclear and fossil fuel plants. To minimize or preclude any unnecessary and possibly adverse effects before and after completion, or during the operation of fossil fuel, thermal, and nuclear power plants, the following general environmental protection measures have been adopted by Taipower:

1. At the very beginning of project planning, the quality of the environment at each site is investigated and a monitoring system set up. At the same time, surveys on ecological, socioeconomic, and scenic resources are carried out so that an environmental impact assessment study can be completed to predict the impact of plant construction and operation and the best practicable solution found to alleviate any adverse impact.

2. New power plants should be built far from population and economic centers and located on the leeward coast.

3. Hydroelectric thermal studies, including mathematical and hydraulic modeling, are carried out to help minimize thermal pollution of marine ecology.

4. Collaborative utilization of natural resources is achieved by establishing a reasonable relationship between landscape planning, power facility construction, and the surrounding areas.

KEPCO is also conducting environmental impact studies simultaneously with its feasibility studies for power plants proposed since 1982. It can be expected that environmental concerns will become increasingly important in fuel choice and siting, as they have been in Japan since the early 1970s.

The planning of transmission facilities is usually constrained by the location of the load and the potential sites available for the power plants. Both KEPCO and Taipower perform load flow and stability studies to analyze the adequacy of their systems. KEPCO now uses mostly Philadelphia Electric Company (PECO) programs for transmission system planning studies and is considering the use of other advanced system planning packages (e.g., an interactive system planning simulator). In transmission system planning, Taipower uses a large-scale power system simulation package (PSS/E), which is a dynamic simulation model that includes synchronous machines, excitation system, governor/turbine, load models, network monitoring models, and DC line models. In the case of planning the distribution system, the complexity of the grids involved forces the use of ad hoc procedures and heuristic planning methods associated with the load flow analysis. Although some commercially available programs try to optimize the integration of generation and transmission, in practice they cannot be used with confidence because of the large number of constraints affecting both generation and transmission system expansions. Consequently, in KEPCO and Taipower integrated generation and transmission system planning is attempted through heuristic approaches of a sequential nature.

RECOMMENDATIONS: INVESTMENT DECISION UNDER UNCERTAINTY

The investment decision determining the size and mix of electrical generating facilities has become extremely difficult and has thus become a challenging task for utility management. This is true not only because the economic life of power plants is about 30 years and the lead time for investment is long, extending up to 10 years, but also because changes in demand, fuel prices, capital costs, and technical performances can drastically alter the relative merits of alternative generation technologies. Since it is extremely difficult, if not impossible, to predict these changes in advance, it is imperative to maximize flexibility in electricity planning to meet the highly uncertain demand 10 years or so hence. Accordingly, the projected configuration of future electrical generating facilities will undoubtedly be revised extensively as additional information makes the projections clearer.

In an effort to cope with demand resource uncertainties, Japanese utilities, KEPCO, and Taipower routinely revise their annual plans, carry out sensitivity analyses, and prepare contingency plans. Experience has shown however, that it is important to perform a systematic evaluation of the uncertainty associated with variations in demand, resources, and technology and the sensitivity of final decisions to the uncertainty of each of the major factors.

Experience has also shown that in view of the possibility of lower growth and structural changes in demand, additional capacity expansions of generating facilities with long lead times, such as nuclear power plants, should be based on relatively conservative forecasts. These should account for the possibility of new baseload demand created by special sales promotions designed to absorb the excess baseload capacity. If higher load growth should materialize, there would still be sufficient time to build coal baseload plants, which require a much shorter lead time for investment. This is particularly true if contingency plans, such as site selection and preliminary design, are made in advance. If future fuel prices are stabilized but capital costs continue to increase, the cost advantage of nuclear generation over coal baseload generation could erode quickly.

Since high-quality construction and operating personnel and procedures are essential to ensure satisfactory performance of nuclear plants, KEPCO is considering scaling down its nuclear program and gradually switching to a less problematic generating technology such as coal. On the other hand, considering the present soft conditions of the world oil market and the absence of any symptoms of a sharp decrease in coal prices in the coming year, it would be a prudent decision to diversify fossil fuel sources between oil, LNG, and coal, with the degree of diversifications depending on the weight given to the security cost of fuel supplies. In so doing, the electric utilities should also take into due consideration the social cost of future surplus residual fuel oils.

KEPCO's total construction costs could be reduced 10–15 percent by locating additional nuclear units of the same type on an existing site. Similarly, significant capital outlays for coal handling, ash disposal, and sea freight can be saved by installing additional coal-fired plants on existing sites. In addition to the cost factors, KEPCO considers future load distribution and environmental constraints in determining the location and size of the power plants and then follows with the transmission and distribution system expansion plan. This semi-integrated approach appears to be realistic because it typically takes no longer than two to three years to construct the transmission and distribution facilities and because changes in population distribution and the location of new industrial complexes over this period can be easily predicted. There may be difficult environmental problems created, however, by siting a number of coal or nuclear power plants at the same location.

Table 4.22
Transmission and Distribution System Expansion Plans

	KEPCO			Taipower		
Item	1981	1986	1991	1981	1986	1991
Transmission (c-km)						
345 kV	2,100[a]	4,200	6,500	1,100	2,300	2,600
154 kV and below	11,000	13,600	16,400	NA	NA	NA
161 kV and 69 kV	NA	NA	NA	7,200	9,000	11,100
Total	13,100	17,800	22,800	8,300	11,300	13,700
Substation (MVA)						
345 kV	7,300	16,300	27,800	6,000	11,500	13,500
154 kV and below	13,700	21,000	31,100	NA	NA	NA
161 kV and 69 kV	NA	NA	NA	18,800	28,200	36,600
Total	21,100	37,500	59,000	24,800	39,700	50,100
Distribution (km)						
Line length	129,200	171,300	229,600	110,000	142,900	204,300

Sources: Ke et al. (1983); Kim et al. (1983).

a. Numbers have been rounded to the nearest hundred.

NA Not applicable.

KEPCO's and Taipower's long-term transmission and distribution system expansion plans are shown in Table 4.22. The expansion rate of KEPCO transmission, substation, and distribution facilities appears quite stable in the period 1982–1991. On the other hand, the expansion rate of Taipower 345kV transmission and substation facilities in the period 1987–1991 appears much lower than the period 1982–1986, because a pessimistic load forecast was made in 1982 and used for generation and transmission and distribution expansion planning. In the 1980s, all major new construction projects were slowed down, but the main projects under construction were in commercial operation by 1986. From 1987 to 1991, few generation units are planned. Consequently only a small number of transmission and distribution facilities will be needed.

CONCLUSION

The power plants, substations, and transmission lines are the physical representations of a power utility. Their maintenance and operation are what occupy the daily routine of its personnel. The planning of these supply systems so as to achieve power at the lowest possible cost with reliability over the long term, however, is a task requiring the detailed consideration of a wide range of factors beyond those related simply to the technologies themselves. In addition, unlike the factors that determine daily operations, many of the factors important to planning cannot be predicted or otherwise determined with accuracy. Probabilistic thinking and methods are needed to enable the decision makers to find the optimum path. Finally, many of the final judgments will not be made entirely on technical and economic criteria. Social and political weights play significant roles, particularly regarding the value of supply security.

Among supply technologies, such considerations are perhaps most dominant in nuclear power. In addition, the utilities of Northeast Asia have chosen this technology to provide the bulk of their expansion in the coming years. For these reasons, it is appropriate that the next chapter examines in some detail the role of nuclear power in the region.

REFERENCES

Clark, E. L. (1986). "Cogeneration—Efficient Energy Source." *Annual Review of Energy* 11:275–294.

"Commissioning of Japan's Futtsu LNG combined-cycle power plant" (1987a). *International Power Generation* 10(4):30–32.

"Korea—Tidal power feasibility study" (1987b). *International Power Generation* 10(6):18.

Marshall, E. (1984). "The procrastinator's power source." *Science* 224:268–270.

CHAPTER 5

NUCLEAR POWER

K. R. Smith and D. J. Rose

The early promise of nuclear power's role in economic development has not yet been realized on a global basis. Since the first oil shock in 1974, only three more developing countries have been added to the original list of three with nuclear power. Although fourth and fifth in this historical sequence, China (Taipei) and Korea have far surpassed all other developing countries in developing nuclear capacity (Table 5.1). Indeed, it could be argued that these are the only two developing countries that have been able to successfully integrate nuclear power into development efforts. In the same period, Japan built the world's largest nuclear power industry in a nonnuclear weapon state.

In addition to total installed capacity, as Table 5.2 shows, there are several other ways to compare nuclear power programs in different countries. Thus, according to number of reactors, Japan places fifth in the world hierarchy and China (Taipei) and Korea about twelfth and eighteenth, respectively. By percentage of total power produced, however, the nuclear programs of France and China (Taipei) stand out. When considering the role of nuclear power in development, it is useful to examine still another indicator for nuclear programs, as shown in the last columns of Table 5.2. These list the nuclear economic intensity of each country—that is, how many nuclear megawatts does the country have installed or under construction per billion dollars of national income? By this indicator, France is again high but not more so than Korea, and neither is as financially committed to nuclear power as China (Taipei). No other developing country even comes close.

Although these indicators vary somewhat according to year, currency exchange rates and other variables, the pattern seems clear. In the 1970s, Korea and China (Taipei) embarked on a path of intensive nuclear power development that seems, at least on the surface, to follow that of Japan.

Although the government and utility orientation toward nuclear power development in the three countries have similarities, there are significant differences in implementation. Basically, however, each country views nuclear power to be one of the principal ways to achieve a greater degree of energy security. Compared to fossil fuel cycles, the uranium fuel cycle is not significantly constrained by physical resource limitations at either global or regional scales. It has been greatly constrained in some countries, however, by failures to develop the technological, managerial, and consensus-generating "resources" that nuclear power requires in such abundance. The Northeast Asian countries have shown that they possess considerable comparative advantages in rapidly developing sustained and efficient utilization of these human resources.[1]

Table 5.1
Nuclear Power in LDCs According to Rated Capacity as of Mid-1986

Country	Year of First Operation	Percent of Total LDC
In operation or under construction		
India	1969	10
Pakistan	1972	1
Argentina	1974	7
Korea	1978	33
China (Taipei)	1978	22
Brazil	1984	8
Mexico	(1988)	6
Cuba	(1989)	4
China (Beijing)	(1990)	9
Total		100
Most likely by 2000		
Turkey		
Egypt		
Possible but unlikely by 2000		
Philippines and Iran, which have partially completed units in place		
Bangladesh, Indonesia, Libya, and Thailand		

Sources: Compiled from Masters (1986) and the world surveys in *Nuclear Engineering International* (August 1986) and *Nuclear News* (February 1985, February 1986).

Notes: Less developed countries (LDCs) are defined here to be non-European nations with less than the world mean per capita income (US$2,900 in 1984). China (Taipei) and Korea account for a large fraction of total LDC nuclear power, but only about 3 percent of the global capacity, which was nearly 400 GW, 37 percent of which was under contruction at that time. This list does not include any Eastern European countries (most of which have nuclear power programs), although several might be considered LDCs by some definitions. Nor does it include South Africa, which started the first of its two 920 MW units in 1984.

Table 5.2
Global Comparison of National Commitments to Nuclear Power

Country	Number of Operating Units[a]	Percent of Electricity Generated[a]	Economic Intensity[b]	
			Operating	Under Construction
Korea[c]	4	26	26	76
Spain	8	24	15	63
France	43	65	50	57
Hungary	2	24	18	53
China (Taipei)	6	53	80	47
Czechoslovakia	5	15	13	32
Germany, Federal Republic of	20	31	16	31
Canada	16	13	24	27
Belgium	8	60	44	26
Switzerland	5	40	20	31
Sweden	12	42	86	25
South Africa	2	4	0	25
United States	101	16	20	20
Philippines	0	0	0	19
U.S.S.R.	50	10	17	18
Japan	33	22	15	14
United Kingdom	38	19	20	12
Argentina	2	10	14	10
Brazil	1	2	2	10
India	6	2	5	8
Mexico	0	0	0	7
China (Beijing)	0	0	0	1
Italy	3	4	4	6
Finland	4	38	49	0
Bulgaria	4	32	43	0
Yugoslavia	1	5	8	0
Netherlands	2	1	4	0
Pakistan	1	1	4	0

Notes: Compared here are number of units, percent of total electricity generated by nuclear plants, and the economic intensity as indicated by the nuclear capacity per billion dollars of national income. The countries are ranked according to the economic intensity of nuclear plants under construction. See Tables 5.4, 5.5, and 5.6 for more details of the programs in Japan, China (Taipei), and Korea.

a. In early 1986; from Masters (1986).

b. MW per billion U.S. dollars of GNP in 1983. Nuclear power plants operating or under construction as of December 30, 1983. From "World List of Nuclear Power Plant," *Nuclear News* (February 1984), pp. 71–90; *International Financial Statistics* (Washington, D.C.: International Monetary Fund, September 1984); *Handbook of International Trade and Development Statistics 1983* (New York: UNCTAD, 1983); *World Development Report 1984* (New York: Oxford University Press [for the World Bank], 1984).

c. In 1986 six units supplied 44 percent of electricity (Masters, 1987).

HISTORY OF NUCLEAR POWER IN THE REGION

After a decade in which all research and development in nuclear matters were prohibited, the Japanese diet established the Atomic Energy Commission in 1956. The first power reactor in Asia was a British natural uranium, gas-cooled reactor, the 160 MW Magnox, ordered by Japan in 1960 and put into operation at the Tokai Mura research and development center in 1966. This sale startled the U.S. nuclear industry, which had expected to dominate the export market to Japan. Consequently, the U.S. government and industry worked hard to establish a strong position in the Japanese market through long-term enriched fuel contracts and offers of technology transfer. By the mid-1960s, the Japanese had decided to rely solely on U.S. light water reactors (LWRs) for the immediate future, partly because of the changes in the U.S. position and partly because of the uncertain future of the British reactor export industry and the technical problems of the Magnox reactor. But even so, for many years the Tokai Mura Magnox consistently outperformed Japan's U.S.-style LWRs, although the differences in age and size make comparison questionable.

At the end of 1962, it was decided that the three largest Japanese utilities—Tokyo, Kansai, and Chubu—should each build a reactor, carry out technical training, and develop long-term plans for nuclear power as a component of their systems. The other six utilities were to build reactors only if the economics and reliability of nuclear power in Japan could be established. As Table 5.3 shows, except for the 440 MW boiling water reactor (BWR), Shimane, ordered by Chusoka Electric Power Company in 1967, this plan was followed in all orders up to 1970. Both BWRs and pressurized water reactors (PWRs)[2] have been built in Japan, but each utility has stuck to one type, although sometimes siting its plants in another utility's territory.

Table 5.4 shows that Japanese nuclear power development proceeded steadily but not rapidly during the early years. By the end of 1974 there were just eight plants in operation, an average annual addition of about 430 MW since 1966. On average during this period the Japanese economy grew by an additional 8.4 trillion yen (all currencies stated at 1975 prices and exchange rates), and the economy absorbed nuclear power plants at the rate of one MW per US$63 million (19 billion yen) of increased gross national product (GNP). Between 1975 and the end of 1983, however, more than one GW was added each year at an absorption rate of about one MW per US$27 million (8 billion yen). Assuming that nuclear power cost about US$500/kW in the earlier period and US$1,000/kW in the later period, these absorption rates indicate that less than 1 percent of the annual GNP increment during the earlier period and 4 percent during the post-oil shock period was devoted each year to nuclear power plants.

Although China (Taipei) and Korea started their nuclear power programs 12 years later than Japan (Tables 5.5 and 5.6), they now have the most aggressive nuclear power programs in the world. Korea's commitment to nuclear power expansion in early 1984 was nearly three times that among countries below the mean global GNP per capita (approximately US$2,900). China (Taipei) has a somewhat smaller construction program, but its total program is the largest in the world and almost five times that of any other country below the mean per capita GNP. These expansion plans may be difficult to finance, not only because of the capital limitations of lending institutions but also because of the magnitude of the loans in relation to the national economies. The total marginal cost of the 76 MW per US$1.0 billion under construction in Korea probably represents nearly 15 percent of one year's GNP. The absorption rates of nuclear power in its neighbors' economies were much higher during the first years of development than those in Japan during a comparable period. In China (Taipei) and Korea, the annual rates of growth of nuclear power during the first six years (1978–1983) were 540 and 310 MW, respectively. These represented absorption rates of one MW for approximately US$2.6 million (0.1 billion NT$) and US$3.7 million (1.8 billion won), respectively. This means that about 40 percent and 25 percent, respectively, of each year's additional GNP were devoted to nuclear power projects (at US$1,000/kW).

Briefly, in the early 1980s, Korea had plans for an even larger commitment to nuclear power. In 1981, however, it announced an indefinite delay in construction of Korean nuclear units (KNU) 13 and 14, and in 1983 a delay in construction of KNU 11 and 12 in the third revision of the 1982–1986 plan. Part of the reason given by the government was the difficulty of raising the needed capital when the country already had a foreign debt of nearly US$40 billion (NEI, 1/84, p. 9). In spite of these changes, Korea has been one of the few countries to keep reasonably close to the nuclear projections of the early 1970s. In 1973, for example, the International Atomic Energy Agency (IAEA) projected that Korea would have 9,800 MW in 1990, which was double the official Korean plans at the time. Before delaying KNU 11 and 12, it seemed possible that Korea might have 9,400 MW on-line in 1991, as Table 5.7 shows, but the total will actually be about 7,500 MW (Payne, 1987). Indeed, although there are plans for five more units before the year 2000, only two have been ordered (Masters, 1987). Thus, unless new contracts are negotiated by 1989, it appears that Korea's schedule will have slipped by almost a decade since the forecasts of the early 1980s.

In contrast to many other countries, China (Taipei)'s electrical demand during the 1970s grew substantially faster than had been projected in 1970. Taipower planned to expand nuclear capacity during the rest of the century to provide most of the new electrical capacity. As did both of its neighbors, however, China (Taipei) scaled down its goals in the early 1980s. In order to reach the former targets set for the year 2000, it would have been necessary to build about one reactor per year. Plans announced in 1983, however, show

Table 5.3
Evolution of Japanese Nuclear Industrial Capacity

				National Base of Major Plant Contractors and Suppliers					
Years of Order	Years of Operation	Size in MW/Type	Utility	Main Contractor	Architect-Engineer	Reactor Manu-facturer	Fuel Contractor	Turbines	Comments
1960	1966	160 Magnox	JAPCO	UK	UK	UK	UK	UK	Turnkey; carbon dioxide cooled, graphite moderated reactor
1965	1970	342/BWR[a]	JAPCO	US	US	US	US	US/J	Turnkey
1966	1971	440/BWR	Tokyo	US	US	US	US/J	US/J	Turnkey
1966	1970	320/PWR[b]	Kansai	US/J	J/US	US	US/J	J	
1966	1972	470/PWR	Kansai	J	J	J	US	J	
1967	1973	760/BWR	Tokyo	J/US	US	US/J	J	US/J	Step increase in BWR size
1967	1974	440/BWR	Chusoka	J	J	J	J	J	Shimane, first all-Japanese nuclear power plant
1967	1976	515/BWR	Chubu	J	J	J	J	J	Hamaoka-1, first all-Japanese turnkey nuclear power plant
1968	1979	165/Fugen	PNC	E/J	—	E	J	J	Prototype light water cooled, heavy water moderated reactor
1969	1974	780/PWR	Kansai	US	J/US	US/J	US	J	Step increase in PWR size
1969	1975	780/PWR	Kansai	J	J	J	US	J	

1970	1983	500/BWR	Tohoku	J	J	J	J	J	
1970	1975	529/PWR	Kyushu	J	J	J	J	J	Turnkey
1970	1976	750/BWR	Tokyo	J	J	J	US	J	
1970	1978	814/BWR	Chubu	—	J	J/US	J	J	
1971	1979	1120/PWR	Kansai	US/J	US/J	US	US	J	Step increase in PWR size
1971	1979	1120/PWR	Kansai	J	US/J	US	US	J	
1971	1978	760/BWR	Tokyo	J	J	J	J	J	
1971	1978	760/BWR	Tokyo	J	J	J	J	J	
1971	1977	538/PWR	Shikoku	J	J	J	J	J	Turnkey
1971	1978	1056/BWR	JAPCO	US	US	US/J	J	US	Tokai-2, step increase in BWR size and last U.S. turnkey project in Japan
1972	1976	781/PWR	Kansai	J	J	J	US	J	
1972	1979	1056/BWR	Tokyo	US/J	US	US/J	US/J	US	Fukushima Daiichi-6, last LWR with significant U.S. participation
1975	1982	538/PWR	Shikoku	J	J	J	J	J	
1976	1982	1067/BWR	Tokyo	J	J	J	J	J	Fukushima Daina-1, first all-Japanese project over 1 GW

Sources: "World List of Nuclear Power Plants," *Nuclear News* (February 1981), pp. 75–94; "Power Reactors," *Nuclear Engineering International* (July-August 1980), pp. 64–72; *Petroleum News* (November 1980), pp. 14–15.

Note: For all power plants ordered after 1972, there has been no significant foreign participation except for license fees.

a. Boiling water reactor. b. Pressurized water reactor. J Japan; US United States; UK United Kingdom; E Europe.

Table 5.4
Nuclear Power Plants in Japan (mid-1987)

Power Plant	Utility	Reactor Type[a]	Capacity Gross (MW)	Authorization by EPRDCC	Construction Permit	Start of Operation
In Operation						
Tokai-1	JAPCO	GCR	166	Dec. 1959	Dec. 1959	Jul. 1966
Tsuruga	JAPCO	BWR	357	May 1965	Apr. 1966	Mar. 1970
Muhama-1	Kansai	PWR	340	Apr. 1966	Dec. 1966	Nov. 1970
Fukushima I-1	Tokyo	BWR	460	Apr. 1966	Dec. 1966	Mar. 1971
Muhama-2	Kansai	PWR	500	Dec. 1967	May 1968	Jul. 1972
Shimane-1	Chugoku	BWR	460	May 1969	Nov. 1969	Mar. 1974
Fukushima I-2	Tokyo	BWR	84	Dec. 1967	Mar. 1968	Jul. 1974
Takahama-1	Kansai	PWR	826	May 1969	Dec. 1969	Nov. 1974
Genkai-1	Kyushu	PWR	559	May 1970	Dec. 1970	Oct. 1975
Takahama-2	Kansai	PWR	826	May 1970	Nov. 1970	Nov. 1975
Fukushima I-3	Tokyo	BWR	784	May 1969	Jan. 1970	Mar. 1976
Hamaoka-1	Chubu	BWR	540	May 1969	Dec. 1970	Mar. 1976
Muhama-3	Kansai	PWR	826	Jun. 1971	Mar. 1972	Dec. 1976
Ikata-1	Shikoku	PWR	566	Feb. 1972	Nov. 1972	Sep. 1977
Fukushima I-5	Tokyo	BWR	784	Feb. 1971	Sep. 1971	Apr. 1978
Fukushima I-4	Tokyo	BWR	784	Jun. 1971	Jan. 1972	Oct. 1978
Tokai-2	JAPCO	BWR	1,100	Dec. 1971	Dec. 1972	Nov. 1978
Hamaoka-2	Chubu	BWR	840	Feb. 1972	Jun. 1973	Nov. 1978
Ohi-1	Kansai	PWR	1,175	Oct. 1970	Jul. 1972	Mar. 1979
Fukushima I-6	Tokyo	BWR	1,100	Dec. 1971	Dec. 1972	Oct. 1979
Ohi-2	Kansai	PWR	1,175	Oct. 1970	Jul. 1972	Dec. 1979
Genkai-2	Kyushu	PWR	599	Jul. 1974	Jan. 1976	Mar. 1981
Ikata-2	Shikoku	PWR	566	Mar. 1975	Mar. 1977	Mar. 1982

Fukushima II-1	Tokyo	BWR	1,100	Jun. 1972	Apr. 1974	Apr. 1982
Fukushima II-2	Tokyo	BWR	1,100	Mar. 1975	Jun. 1978	Jan. 1984
Sendai-1	Kansai	PWR	870	Mar. 1976	Dec. 1977	Jul. 1984
Onagawa-1	Tohoku	BWR	524	May 1970	Dec. 1970	Jan. 1985
Takahama-3	Kansai	PWR	870	Mar. 1978	Aug. 1980	Feb. 1985
Takahama-4	Kansai	PWR	870	Mar. 1978	Aug. 1980	Feb. 1985
Fukushima II-3	Tokyo	BWR	1,100	Mar. 1977	Aug. 1980	Jul. 1985
Kashiwazaki-Kariwa-1	Tokyo	BWR	1,100	Jul. 1974	Sep. 1977	Oct. 1985
Tsuruga-2	JAPCO	PWR	1,160	Dec. 1978	Jan. 1982	Feb. 1987
Fukushima II-4	Tokyo	BWR	1,100	Jul. 1978	Aug. 1980	Sep. 1987
Hamaoka-3	Chubu	BWR	1,100	Oct. 1978	Nov. 1981	Sep. 1987
Subtotal		34 units	27,880			
Under Construction						
Shimane-2	Chugoku	BWR	820	Mar. 1981		(1989)
Tomari-1	Hokkaido	PWR	579	Mar. 1982		(1989)
Kashiwazaki-Kariwa-2	Tokyo	BWR	1,100	Mar. 1981		(1990)
Kashiwazaki-Kariwa-5	Tokyo	BWR	1,100	Mar. 1981		(1990)
Tomari-2	Hokkaido	PWR	579	Mar. 1982		(1990)
Ikata	Shikoku	PWR	870	Jan. 1982		(1991)
Genkai-3	Kyushu	PWR	1,180	Sep. 1982		(1993)
Genkai-4	Kyushu	PWR	1,180	Sep. 1982		(1994)
Subtotal		8 units	7,410			
Planned		19 units (75% BWR)	20,600	(All to start before 2000)		
Grand Total		61 units	55,930			

Sources: TEPCO and Masters (1986, 1987).

a. GCR gas-cooled reactor; BWR boiling water reactor; PWR pressurized water reactor.

Table 5.5
Major Features of the Taipower Nuclear Power Program

	Nuclear Steam Supply System			Turbine Generators			Construction Period			Construction Costs	
Unit	Supplier[a]	Type	Rating MWth	Supplier	Rating MW	Architect/ Engineer	First Concrete	Fuel Loading	No. of Months	Total (US$ MM)	US$/KW
I-1	GE	BWR4	1,775	W	636	EBASCO	6/72	10/77	64.0	336	470
I-2	GE	BWR4	1,775	W	636	EBASCO	12/73	10/78	58.5	310	
II-1	GE	BWR6	2,894	W	985	Bechtel	11/75	1/81	61.5	1,682	786
II-2	GE	BWR6	2,894	W	985	Bechtel	3/76	3/82	72.0		
III-1	W	PWR	2,785	GE	951	Bechtel	8/78	3/84	67.0	2,436	1,209
III-2	W	PWR	2,785	GE	951	Bechtel	2/79	12/84	69.0		

Sources: Taipower and Masters (1986, 1987).

Notes: As of mid-1987, no bids had been solicited for additional units.

a. GE General Electric (U.S.A.)
W Westinghouse.

Table 5.6
The Korean Nuclear Power Program

Plant Name	Site	Capacity (MW Gross)	Reactor Type	Scheduled Operation	Status	Supplier and Architect/Engineer[a]
Nuclear Unit No. 1	Kori	587	PWR	Apr. 1978	In operation	NSSS, fuel: Westinghouse T/G: GEC A/E: GAI
Nuclear Unit No. 2	Kori	650	PWR	Apr. 1983	In operation	NSSS, fuel: Westinghouse T/G: GEC A/E: GAI
Nuclear Unit No. 3	Wolsung	678	PHWR	Dec. 1982	In operation	NSSS, fuel: AECL T/G: Parsons A/E: CANATOM Co.
Nuclear Units Nos. 5 and 6	Kori	950 950	PWR PWR	Sep. 1985 Apr. 1986	In operation	NSSS, fuel: Westinghouse T/G: GEC A/E: Bechtel
Nuclear Units Nos. 7 and 8	Yeonggwang	950 960	PWR PWR	Aug. 1986 Jul. 1987	In operation	NSSS, fuel: Westinghouse T/G: Westinghouse A/E: Bechtel
Nuclear Units Nos. 9 and 10	Uljin	950 950	PWR PWR	(Sep. 1988) (Sep. 1989)	Under construction	NSSS, fuel: Framatone, Cogema T/G: Alsthom Atlantique A/E (Consultant): EBASCO
Nuclear Units Nos. 11 and 12	Yeonggwang	950 950	PWR PWR	(1995) (1996)	Contracts let in late 1986	NSSS, fuel: KHIC, C-E T/G: KHIC, GEC A/E: KOPEC, Sargent & Lundy

Sources: KEPCO, Masters (1987), and Payne (1987).

a. NSSS nuclear steam supply system; T/G turbine generators; A/E architect/engineer; GEC General Electric of UK; GAI General Atomic, International; AECL Atomic Energy of Canada, Ltd.; C-E combustion-engineering.

a target of 5,430 MW by 1990 and 10,100 MW by the year 2000. After the six plants shown in Table 5.5, the next two (station 4) would not be on-line until 1994 (NEI, 7/84, p. 7). But then, as the sixth unit was put into commercial operation, station 4 (units 7 and 8) was put on hold, partly because of government delays and partly because of rising public concern over safety (NEI, 5/87, p. 14). Thus, Taipower's operable nuclear capacity will remain fixed at 5,100 MW (gross) and 4,900 MW (net) at least into the late 1990s, reflecting a lower growth rate for electricity demand and an increased reliance on coal (Masters, 1987).

In late 1983, Japan scaled down official forecasts for the entire energy sector including power. Nuclear capacity in 1990 was predicted in 1972 as 100 GW, but dropped to 52 GW in 1979, 46 GW in 1982, and 35 GW in 1983. The predicted nuclear power percentage of electricity generation (30–34 percent) and total energy (11 percent) remained the same, however. This was some three GW higher than the total in service and under construction in 1987 (see Table 5.4). For the year 2000, the 1982 forecasts called for 90 GW, but by 1986 they had been reduced to 62 GW, or about 40 and 15 percent of electricity and total, respectively. By 1987, official forecasts had been reduced to 53 GW, representing plants with tentative schedules (Masters, 1987).

Table 5.7
Nuclear Capacities Projected in 1984 for Selected Years (GW)

Year	Japan[a]	Korea[b]	China (Taipei)[c]
1980	15.5 (12)[d]		
1981		0.59 (6)	
1982			3.2 (28)
1986		4.80 (27)	5.1 (30)
1990	46.0 (22)		5.1 (30)
1991		9.40 (40)	
1995		15.8 (46)	7.1 (36)
2000	90.0 (30)	24.00 (49)	10.1 (41)

a. Table 5.4 indicates that about 32 GW could be in service by the end of 1990. Official 1987 projections call for 53 GW by 2000 (Masters, 1987).

b. If units 9 and 10 start as scheduled, Korea will have about 7.5 GW by 1990. Unless new plants are ordered by 1989, however, only about 9.4 GW will be available by 2000 (Payne, 1987).

c. By 1987 it was clear that 7.1 GW could not be reached until the late 1990s, making about 28% of total capacity (Masters, 1987).

d. % of total electric capacity in parentheses.

PERFORMANCE IMPROVEMENTS IN EXISTING AND NEW REACTORS

Taipower's nuclear program has seemingly run into few serious performance problems. The construction times of its four BWRs were shorter than those for similar reactors in the United States and, at about five years, were some of the most rapid construction times in the world. Capacity factors have suffered from the excess of baseload in the Taipower system but seemingly not from technical or managerial problems at the plants themselves. In general, unlike many of the world's reactors, Taipower's plants have improved their performance over time (Chu, 1984). The mean cumulative capacity factor as of early 1984 was about 60 percent, only slightly below the global average for BWRs (NEI, 7/84).

In contrast to China (Taipei), which has diversified into both types of LWRs, Korea has chosen to remain with the PWR. This is the result of a desire not to spread technical personnel any thinner than necessary by trying to develop expertise in all three major reactor types being exported today (PWRs, BWRs, and heavy water reactors or HWRs). Accordingly, the Korean decision to build Wolsung 1 (KNU-3), a 630 MW CANDU,[3] may now seem overambitious, although its construction and commissioning times were exemplary (Bredahl and Phillips, NEI, 9/83, pp. 25–28). It does, however, offer Korea and the rest of the world a unique opportunity to compare the relative performance of an LWR (KNU-2) and an HWR (KNU-3) of roughly the same size (679 MW and 650 MW) and age (July and April, 1983) in the same country (Jeffs, NEI, 3/83, pp. 49–52).[4]

In contrast to the programs of its neighbors, which began much later, the Japanese light water reactors in the 1970s were plagued with a series of minor but debilitating operational problems that led to quite low capacity factors. Clogging of coolant channels, stress corrosion of pipes, cracked pump impellers, fuel pin leaks, and abnormal turbine wear are examples of these problems, which may in hindsight be partly due to uncritical technology transfer (NEI, 7/80, p. 7).

In addition, the plant installation time increased rapidly. A Japanese study in 1980 indicated that once Japanese utilities initiated the process, they averaged 15 years from the first siting proposal to local governments through to operation. This figure compares with an average of 10 years only a few years earlier (Lester, 1982). Construction delays, however, were not the major reason, increasing only from 4.5 to 6 years (NW, 3/13/80, p. 15).

In the mid-1970s, the government embarked on a three-phase program to standardize the design, licensing, and construction of light water reactors and to implement other improvements to increase reliability and safety and to decrease the cost and time of construction. By the end of the first two phases in 1982, countrywide capacity factors had improved to over 70 percent (NEI, 7/84, p. 6), and there was evidence that new plants would be able to sustain

75 percent capacity factors (NEI, 4/84, p. 14). The principal ways in which capacity factors were so dramatically improved were: (1) pinpointing of particular generic technical problems such as wall thinning in the steam generator tubes, improving the materials, and applying techniques to circumvent the problem, (2) shortening the duration of periodic inspections, (3) extending the operating cycle time between inspections, and (4) rearranging equipment within and enlarging the containment building (Murata, 1984).

Part of this program was designed to decrease radiation exposure to workers during routine maintenance and refueling. This was in response to disturbing trends common in LWRs in many countries of increasing occupational radiation exposure as plants aged. Unchecked, these trends can lead to large economic and reliability impacts because of the personnel changes that must be made in order to meet radiation exposure regulations. There is evidence that the Japanese may have been able to reduce exposures in their new LWRs by 50 percent by such measures as rationalizing the equipment layout and developing remote manipulators (NEI, 4/84, p. 21).

To decrease installation times, the Ministry of International Trade and Industry (MITI) established a schedule of standardization for LWRs written not only to stabilize size and design characteristics but to incorporate Japanese modifications to those U.S. BWR and PWR designs and operational procedures that the Japanese believe led to their past troubles (Toyoto and Itoh, NEI, 12/79, pp. 62–65). Part of the standardization effort has been funded by a national tax on electric power, which in 1979 amounted to 0.4 mill per kilowatt-hour (kWh) (Murata, NEI, 12/79, pp. 68–70). Additional portions of these tax revenues subsidize low interest loans to utilities that purchase nuclear equipment made in Japan, and to facilitate siting.

The Tokyo Electric Power Co. (TEPCO) has been one of the leaders in implementing measures designed to decrease construction times and costs of its BWRs. Units 2 and 5 of its Kashiwazaki-Kariwa site, scheduled for 1990 completion, have incorporated such measures as prefabricated equipment, a temporary dome so that work can proceed under all weather conditions, and a large crane outside the containment building. These measures, combined with a greater degree of standardization, reduced construction times and costs. It is estimated, for example, that their costs will be 10 percent lower than unit 1 on the same site (Kawahito and Seko, NEI, 8/84, pp. 39–43). Although these particular plants have relatively long construction times because of the seismic and weather conditions at the sites, ongoing projects at other locations seem to be successfully attaining construction times of 50–55 months (Varley, NEI, 6/83, pp. 37–53).

The improvements in capacity factors, construction, costs, and radiation exposure have all contributed to an enviable record of achievement, and one reflecting well on the Japanese ability to mobilize universities, vendors, government, and the utilities in a common effort. The third phase, which began in 1981, is directed toward improving the load-following and uranium utilization capabilities of nuclear power plants and further improving reliability, installation time, and occupational exposure. Advanced reactors with burn-up increasing from 33 to 40 to 50 GWd/ton (higher fuel efficiency) is an important part of this phase. Another part of this phase includes a cooperative agreement between Hitachi, Toshiba, and Fuji Electric, and Kraftwerke Union (KWU) in Germany to construct a standardized PWR, possibly for Tokyo Electric Power Company (Murata, 1984). In bringing the first European entry into the Japanese market, another tradition would be broken, for it would mean that TEPCO would become the first Japanese utility to operate both types of LWRs. Through bilateral and regional exchanges as well as through such efforts as the IAEA Power Reactor Information System (PRIS), it should be possible to apply the experience gained from the Japanese program to other countries, including Korea and China (Taipei) (Skjoldebrand, White, and Qureshi, NEI, 10/83, pp. 31–35).

Partly as a result of their early problems with LWRs, the Japanese have long been considering other reactor types that might fill the gap between LWRs and the breeder. These efforts culminated in the development of the 165 MW Fugen. It is an advanced thermal reactor (ATR) prototype that is cooled by light water but moderated by heavy water and is capable of using a variety of fuel mixtures at relatively high efficiency. Japanese adoption and purchase from Canada of the CANDU heavy water reactor would provide competition with Fugen. The CANDU would offer only a few of the fuel-cycle advantages of an ATR, although it would add flexibility to Japan's present all-LWR system. In 1979 and 1980 the Japanese seemed to be on the verge of purchasing a CANDU, but by early 1981 the relatively good performance of Fugen and the financial impact of the Three-Mile Island accident had apparently persuaded the Japanese to rely on their own design. Japanese utilities, like many others, will be interested in Korea's experience with its CANDU.

Plans are being made to complete the design of a 1,300 MW demonstration A-BWR to be started by about 1990 (Murata, 1984). A 1,300 MW A-PWR would follow. In addition to other improvements, these reactors are expected to have some load-following capability (see section on Load Following below). There was originally substantial controversy within the Japanese nuclear establishment about the need for ATRs in their overall nuclear strategy, which had been to move from LWRs directly to breeder reactors (Industrial Bank of Japan, 1982). Recently, however, with the changes in the world uranium and reprocessing industries, the ATR strategy is looking more attractive. Japan plans to recycle plutonium in its LWRs, especially in its A-PWRs and A-BWRs. The first deliveries of plutonium from Japanese reactor fuel reprocessed in France began in 1984.

In 1980 MITI announced plans to develop 300–700 MWth light water reactors designed both for electricity generation and steam production at 300 °C

(NEI, 8/80, p. 9). Such reactors might not only have many potential applications in Japan if present siting restrictions were made more "flexible," but might also enable Japan to expand its nuclear industry by exporting to developing countries (see section on Economics and Trade below). They might also allow construction to take place at the factory, reducing costs and construction times.

The long-term plan in all three countries has been to proceed from LWRs toward breeder reactors, but the timing has been modified to beyond 2020 (*Nuclear News*, 5/87, p. 52). Several important strategy and policy issues are connected with breeders. First, the once conventional wisdom that the fast breeder reactor was the logical endpoint of nuclear power progress in most countries has come into serious question. Indeed, work on a U.S. prototype has been indefinitely postponed. Fuel reprocessing costs now appear to exceed US$500/kg of heavy metal, and the cost of making uranium-plutonium mixed oxide fuel (MOX) compared to conventional U-235 enriched fuel is not well known, but is likely to be substantial. The breeder reactor itself will cost considerably more than a nonbreeder of the same capacity.

Thus, it appears that natural uranium would have to cost several hundred dollars per kilogram before breeders (for which high fuel efficiency is the only major advantage) would become economically attractive, even disregarding the political difficulties attending the use of plutonium. When will this be? Almost surely not before the year 2020, even with present reserves. It is also possible that uranium may be produced inexpensively from seawater or that new large deposits will be discovered on land. The chance of at least one of these possibilities coming to pass is high enough to warrant serious attention to development of more efficient converter reactors to operate on a once-through fuel cycle rather than plan on any system requiring reprocessing.

It is useful to consider what design converter reactors might have and who could supply them. The U.S. reactor industry, beset by utility overcapacity, public opposition, nonideal regulation, and technical and economic problems, is becoming increasingly stagnant, concentrating more and more on service. A resurgence of U.S. demand is predicted for the early 1990s, but if they persist on their present course, U.S. vendors are unlikely to be prepared for it. In addition, the U.S. domestic market has little need for load-following reactors. Thus the next generation of flexible LWRs is likely to come from non-U.S. sources, very likely Japan or as the result of collaborative or competitive efforts of Japan and Germany. The United States and Japan do cooperate on some developments, but increasingly the initiative is turning to Japan.

Development of even more advanced reactor types is at present speculative. Canadian reactors, both the CANDU and the organically cooled model, might be suitable candidates for developing advanced once-through systems, but the Canadian domestic market is stagnant and little overseas support exists. The HTGR types, such as the German pebble-bed reactor, are interesting for both power and process heat, but there are questions as to whether they can be made to run economically on a once-through fuel cycle. The possibility of "ultrasafe" reactors receives much public attention, but it is not certain how much may be given up in other features such as ease of repairability or maintenance. An interesting long-range opportunity appears to exist, say for the year 2000-plus period, for any country that wishes to take advantage of it. Under present circumstances, however, this is beyond the economic time horizon of the U.S. private sector.

MANAGEMENT ISSUES

International comparison of nuclear power programs shows how critical good management is to success. In addition, nuclear power activities touch upon other important national goals such as localization and education.

Architecture/Engineering Capability

A 1,000 MW nuclear power plant requires the sophisticated integration of some 15–18 million high-quality components. A systematic body of engineering knowledge must be accumulated. In particular, a high level of architecture-engineering (A/E) capability is a prerequisite for the successful construction and operation of a nuclear power plant. In light of the importance of accumulating domestic capability, Korea, China (Taipei), and Japan have been taking some measures that merit special attention.

In the case of China (Taipei), the government has been encouraging closer cooperation between foreign and local engineering firms. Two joint venture engineering companies have been established: Pacific Engineers and Constructors Inc. (PECI) between Local Sinotech Engineering Consultants Inc. and Bechtel; and EBASCO-CTCI Corporation between CTCI and EBASCO.

PECI was established in May 1979 to offer professional engineering, procurement, construction, start-up, and management services internationally. To build its technical capabilities, Bechtel/PECI has established technology transfer programs including both classroom and on-the-job training. With some 450 national engineers, PECI can now perform various services related to power and energy (including nonnuclear) with some Bechtel support.

EBASCO-CTCI was established in May 1980 to provide services and technology transfer in many of the same fields. In the mid-1980s the company developed the capability to undertake various projects and services successfully.

Consequently, a gradual decrease in the foreign consultants' assistance in the nuclear power projects of China (Taipei) will be seen. For future nuclear projects, design work will be done by the local architects-engineers.

In the case of Korea, the A/E services are increasingly being taken over by the Korea Power Engineering Company (KOPEC). As of the end of 1982,

96 percent of its total stock was owned by the Korea Electric Power Corp. (KEPCO), which constitutes a quite different arrangement from the case in China (Taipei). Korea's ultimate objective is to have KEPCO carry out future nuclear power plant projects in association with KOPEC, which will be designated as the prime contractor so that services by foreign architects/engineers will be minimized. At present, however, KEPCO awards various engineering contracts to KOPEC, for which it hires qualified foreign consulting engineers as subcontractors. Thus KOPEC is, among other things, an agent of technology transfer.

Based on turnkey contracts, Korea completed the phase first of its nuclear program with its HWR and first two PWRs. All the later PWRs, however, have had substantial Korean involvement. The first six PWR contracts were awarded to Westinghouse for the primary system and GEC, United Kingdom, for the turbine generator. In 1980 the Koreans apparently felt the need to diversify their supply source, and the four new orders announced that year (KNU 7–10) were split between Westinghouse and the French company Framatome. Although the two French reactors are Westinghouse-style PWRs built under license, by choosing Framatome the Koreans gain access to different financing channels and different fuel suppliers. They also hope to gain access to the considerable French nuclear experience in breeder reactors.

The second phase of Korea's nuclear program will be complete with the completion of units 9 and 10, which has involved KEPCO construction and start-up management, but with foreign architect/engineering services and equipment. In phase three, which began with units 11 and 12, Korean organizations will be the principal contractors for all aspects of design, construction, and start-up (Payne, 1987).

Japan meets the A/E requirement quite differently. Instead of establishing independent A/E companies, giant plant makers such as Hitachi, Toshiba, and Mitsubishi Heavy Industries deal with the entire plant and have expertise in both software and hardware. They have accumulated their own A/E capability within their own structures. Japanese utilities also have extensive engineering capability and share the A/E tasks with the suppliers.

In Japan A/E capability is thus met by close collaboration between the domestic vendors and the utilities. This mode of operation has been very successful, but this vendor-utility collaboration is not unique with Japan. In the United States, for instance, Duke Power Company does its own architecture-engineering, and its performance to date has generally been excellent. Their nuclear plants have been built close to budget and approximately on time.

The different approaches to the A/E tasks reflect, among other factors, the differences of supply in the three countries. China (Taipei) must import most of its nuclear power reactor equipment from abroad, principally from the United States. The U.S. vendors have, on the whole, little experience in the A/E business, geared as they are to the prevalent U.S. practice of electric utilities dealing with A/E firms that in turn deal with a multiplicity of vendors. Thus China (Taipei) must acquire A/E service either by importing the service or establishing domestic capability. It has been steadily moving from importing the service to developing a domestic one.

A similar situation applies in Korea, except that the A/E function is coming to be performed de facto inside KEPCO itself, via its almost wholly owned subsidiary KOPEC, which seems to act as a major division of KEPCO (Payne, 1987). Thus, apart from organizational details, the arrangements in China (Taipei) and Korea are to a large extent determined by the circumstances of foreign equipment supply and domestic desire for independence and localization.

The Japanese arrangement has much to recommend it. A question can be raised about separate A/E companies: Do they act as smooth facilitating matches between vendors and utilities, or do they act as barriers? The objectives of the A/E companies are not necessarily the same as those on either side of them. Whereas they bring broad expertise from many projects to bear on their tasks, this very expertise tends to inhibit development of such expertise by either the vendors or the utilities. Thus utility needs and vendor development may both suffer. Furthermore, the electric utilities tend to become the mere custodians and operators of the equipment, not full partners in the direction of their own future, and sometimes they are unpleasantly surprised at the complexity of what they have bought.

Why do U.S. electric utilities rely for the most part on A/E firms, and Japan does not? One important reason is that most U.S. electric utilities are relatively small compared to the Japanese ones, so each utility company cannot afford to have full A/E capability, especially if the major U.S. vendors have little capability to match. Thus the U.S. arrangement may have less to recommend it when the vendors and electricity utility companies are large enough to bridge the gap—a size that is probably necessary in any event if the electric utility company is to run its nuclear reactors optimally.

All this is not to say that intermediate-size utility companies, with competent and attentive management, cannot work very successfully with vendors and A/E firms to bring nuclear power plants on-line within budget and according to schedule. That happens even now in the United States as well as in East Asia. Management skill and attention seem to be key factors here as in any enterprise.

Localization Plans for Reactor Component Manufacturing

Japan's experience with its Tokai No. 1 reactor, its first and last gas-cooled reactor, has helped the Japanese to place greater emphasis on localization. Thus from nearly the beginning, an important characteristic of U.S. reactor sales to Japan was the high component of technology transfer involved. This set the precedent for the nuclear programs of Korea and China (Taipei), which began much later. Under licensing arrangements with the two largest U.S. reactor manufacturers, Westinghouse and General Electric, several Japanese companies rapidly developed sufficient capability to design, fabricate, and construct complete nuclear power stations. Table 5.3 illustrates this transition. Note that the Japanese practice was to secure direct U.S. industrial participation, particularly when changing to larger models of each type of LWR. The growth in the size of U.S. LWRs, however, seems to have leveled off at about 1,300 MW. The Japanese have had indigenous industrial capacity at near this size since construction began on the 1067 MW Fukushima Daina-1 in 1976. As Table 5.3 indicates, since the three large plants went into operation in 1979, U.S. direct participation in the Japanese market has been much less prominent, although still important in commercial arrangements to help Japanese companies compete within Japan.

For the foreseeable future, U.S. participation in the Japanese market will consist of selling relatively minor technical consulting services and collecting license fees. In spite of the large license fees collected, the U.S. nuclear industry now apparently questions the overall economic wisdom of having helped the Japanese develop industrial capacity so rapidly. Total Japanese reactor construction capacity is about 6 GW per year, although, as in most of the world, this capacity seems likely to be underutilized in the near future, and the three vendors—Hitachi, Toshiba, and Mitsubishi—only began to achieve marginal profits in 1980 (Industrial Bank of Japan, 1982).

The principle of promoting independent development and self-help has enabled Japan to produce almost all components domestically. Imports in 1982 were limited to 39 items for a BWR of the 1100 MW class and 20 items for a PWR. In terms of value, the domestic manufacturing ratio exceeds 90 percent, and almost all of the remainder is not manufactured domestically because it is cheaper to buy them abroad and the scale of production does not at present justify localization. In fact, Japan probably holds comparative advantage in the international market for a number of nuclear products. With its large domestic capability but recent projections of moderate domestic orders, Japan is seeking foreign sales (see section on Nuclear Trade Futures below, p. 178).

In China (Taipei), domestic manufacturers have successfully fabricated certain basic components such as liner plate, mechanical penetrations, low pressure valves, and piping. Taipower is determined to promote localization of such additional components as piping loops, other valves, and heat exchanges. This localization will presumably apply to construction of delayed units 7 and 8. In the long run, both engineering and manufacture of additional components are to be undertaken domestically for units 9 and 10. The components are mainly products needed by other local industries and, sometimes, promotable in foreign markets. Manufacturers will localize technically complicated components only when they obtain technologies through joint ventures or licensing agreements.

Korea has a much more ambitious localization program. Most key components are manufactured locally as a result of technology transfer through license and/or technical assistance agreement up to 1991, with such technical partners as Combustion Engineering, Westinghouse, General Electric, and Framatome. Even after having reduced the dependence on foreign technologies through experience with nuclear units 7 to 10, KEPCO is continuing to make use of foreign technology to ensure product quality and technological development.

The effort in Korea to follow the Japanese example by developing domestic nuclear construction capacity has been major and successful. Labor participation of KOPEC has risen from one-quarter to four-fifths from units KNU 5 and 6 to KNU 11 and 12. Localization plans through the 1980s called for 26 percent for the nuclear steam supply system, 40 percent for turbine/generator, and 55 percent for the balance of the plant. Units 11 and 12 (1995–1996) will involve two-thirds or more Korean contribution in each category. All future plants will probably be 100 percent Korean (Payne, 1987). This ambitious localization plan creates large demands for trained personnel and explains the country's vigorous program of sending engineers abroad for education and training.

Education and Training

This topic, closely related to domestic programs of localization and provision of engineering services, has received much attention in Japan for many years, and now Korea and China (Taipei) also have substantial programs.

Japan's program is quite strong, as might be expected. Some 47,000 professionals were in the nuclear sector work force in 1982, with 7,000–9,000 of them in electric utility companies. Almost all of these were trained domestically. In 1982 Japanese universities were turning out annually 460 bachelor's degrees and 150 master's degrees in science, and 80 Ph.D. nuclear engineers. The Japan Atomic Energy Research Institute (JAERI) also runs training courses. In 1980 there were 20,300 registered nuclear engineers.

The present Korean nuclear regulatory system has been burdened in part by lack of technical personnel. The Nuclear Regulatory Bureau (NRB), which functions as secretariat to the Korea Atomic Energy Commission (KAEC), for example, had only 20 persons with bachelor's degrees in 1980; turnover is high because of low salaries. Part of the problem in attracting competent personnel to the NRB and other parts of the government's nuclear apparatus is extensive competition from private industry. The combined labor requirements of the utility (KEPCO), component manufacturers, research and development (Korea Advanced Energy Research Institute, or KAERI), or the architect and engineering company, KOPEC, and government are very high. Recent analyses indicate that the long-term interests of the country would be better served by lowered objectives in both capacity growth and growth in localization. Such rapid growth may jeopardize safety and reliability as well as distort the financial and labor markets. There is now considerable debate within the South Korean nuclear establishment on the appropriate level of localization.

In response to these pressures, as of 1982 about 2,000 persons had been trained domestically for the Korean nuclear sector and about 500 had been trained overseas, including engineers, operators, and technicians. KEPCO and KOPEC have their own training programs, and a nuclear engineering department has been established at Ulsan University. Overall, about 200 bachelor of science degrees and 20 Ph.D.s in nuclear engineering are awarded annually in Korea (Payne, 1987). The strong Korean localization policy may, however, necessitate an increased need for trained personnel.

In the early 1980s, China (Taipei) locally trained 1,200–1,500 people per year in all categories, plus 630 from overseas, and Taipei University had about 80 graduates a year in the nuclear field.

Simple comparisons of these numbers are instructive. Japan's well-established nuclear industry is well staffed and has about 20 times the number of people in the professional field as do either Korea or China (Taipei). By the year 2000, each of these countries expects to have nuclear-electric programs about 15–20 percent of the size of Japan's. Thus the need for expanded training programs in Korea and China (Taipei) seems urgent.

The architect/engineer, localization, and personnel/training issues are part of a larger whole—the personnel-machine software—and pose critical questions: Who does what? How are the decisions made? Where do particular categories of people go? To what sector do they owe their allegiance? Education and training affect the outcomes. Until recently, for example, front-rank engineers, especially with advanced degrees, were rarely found or seen to be needed in U.S. electric utility companies. Nuclear power systems are high tech, and the advent of these and other advanced power systems—including control and dispatching technologies—in the utilities has changed all that.

LOAD FOLLOWING

All three East Asian countries, like France, plan to rely on nuclear power providing as large a fraction of electric power generation as can be achieved rationally and economically. Naively, one might think that this would mean all the baseload, but that is an oversimplification. First, not all the baseload *should* be nuclear, especially only one type, lest some generic fault forces it all to be shut down at once. Thus at least a modest fraction should be something else—in the present case, coal. Furthermore, most of the large intermediate load plants cannot be shut off completely, but must be kept operating at reduced power to be able to respond to daily variations. Hydropower and turbines are exceptions, but the first is in short supply in the three countries, and the second is fuel-expensive, suitable mainly for peaking.

Nuclear power plants cannot now follow major changes in diurnal load (e.g., those caused by differences between daytime and nighttime demand). A principal reason is that the fuel pins themselves cannot withstand many center-to-surface temperature fluctuations. Another is that the control rods are not conveniently arranged for continuous fine power control. Operation with control rods partly inserted tends to lead to uneven power distribution throughout the reactor core and a resultant inefficient fuel use. Also, at part load, the turbines will have to handle wetter steam, which causes turbine blade damage. Modifying reactor systems to avoid these problems is not impossible, but it would require significant redesign.

The first country to face this problem was China (Taipei), where in 1983 about 30 percent of the installed generating capacity was nuclear. At present, the percentage of nuclear power in the daily total varies from 30 percent to more than 60 percent, and Taipower's difficulty is aggravated by (1) its gradually deteriorating daily load curves, with an increasing peak-to-minimum ratio (see Chapter 3), and (2) an excess of nuclear baseload plants, which were ordered during a period when electricity demand forecasts were higher than those of today. Taipower's short-term resolution of this problem was to select one nuclear reactor to run at a reduced but steady output for an extended period, say a few weeks, and then switch to another reactor. This may have been the best available supply management response, but there may also have been cost-effective ways to make the daily load curve less variable (see Chapter 3).

A future supply management option might be to install reactors capable of following the major diurnal trends far enough up into the intermediate-load region so that some nonnuclear intermediate-load facilities could be eliminated. The detailed optimization of such systems is complicated, but the main idea is straightforward. Because the levelized cost of nuclear power is less than that of intermediate-load power (otherwise nuclear plants make no sense), and because the intermediate-load region starts at 100 percent on the load duration curve, the minimum cost solution consists of overbuilding the nuclear component and running the combination so that the levelized cost of the last increment of part-time nuclear power equals the cost of the nonnuclear power it displaces. Since the delivered price of overseas coal is about US$60/MT or more in Japan and nearly the same in Korea and China (Taipei), the economics may favor some degree of intermediate-load nuclear power if the equipment can be provided. As mentioned above, Japan is active in the development of such reactors. Several of its A(Advanced)-PWR and A-BWR designs, planned for the early 1990s, will have redesigned fuel and core layouts, improved control systems, and other improvements designed for load following. China (Taipei) and Korea would do well to consider this option when they order reactors for the 1990s or later.

SPENT FUEL, REPROCESSING, AND WASTE

As in most countries, after its removal spent fuel is first stored in on-site fuel pools in Northeast Asia. Taipower's nuclear units (1–4) have 11 years of on-site reactor storage available; unit 5 and those built after unit 5 are capable of storing 15 years' worth of spent fuel (Chu, 1984). Future plants will probably have facilities for lifetime spent fuel storage, and additional on-site storage will probably be needed at existing reactors. At present there is no immediate plan for centralized storage or reprocessing. Korea has designed its spent fuel pools to reach capacity by 1995. At that time, the fuel will be sent for reprocessing or to temporary away-from-reactor (AFR) centralized storage sites. Since an AFR is estimated to take 10 years to build, Korea is planning to make its spent fuel decision in the late 1980s (Sung, 1983). In Japan each utility must show that it has made arrangements for storage or disposal of spent fuel over the lifetime of a reactor as part of the licensing process. This is done at present mostly through a combination of on-site storage and shipment to Europe for reprocessing.

Korea and China (Taipei) have yet to announce decisions about reprocessing, direct disposal of spent fuel, or long-term storage in AFR facilities. Part of their reluctance stems from the fact that such schemes work best when associated with large national programs, say 30 GW of nuclear capacity or more. Therefore, they look beyond their borders to Japan, the United States, and Europe for leadership in these matters, even though progress has been slow there as well (NN, 3/87, pp. 44–88). In addition, because plutonium is a sensitive material, it is felt by most observers that an arrangement agreed upon internationally is preferable to any independent national one (Hirata, 1984). These factors plus the dismal prospects for early breeders seem to make fuel reprocessing unlikely in these countries.

Japan reprocesses about 200 tons of spent fuel per year (from six reactors, approximately) in its demonstration-scale facility at Tokai Mura and intends to reprocess most or all of its spent fuel eventually. It has contracts for reprocessing in the United Kingdom and France with most plutonium presently being stored in both countries, along with the fission products. The plutonium is being returned to Japan for experiments in making MOX and/or breeder fuel. By contract, the fission products must be returned to Japan within 25 years; initially they will be stored at Tokai Mura along with wastes generated there. Commercial use of MOX is planned for 1997 (Suetsuna, NEI, 2/87, p. 39).

The combination of European contracts and Tokai Mura reprocessing are handling Japan's needs during the 1980s. At present the plan for the 1990s seems to be to operate a reprocessing plant with capacity of 800 tons per year (1,200 t/y nominal capacity) to be built at a site that will also accommodate other fuel cycle facilities built by the Japan Nuclear Fuel Service Co. (JNFS). The company was established in 1980 and is owned by 100 shareholders among industrial and utility companies (NN, 8/84, p. 77). As already noted, however, such a recycling policy looks increasingly unattractive for the remainder of this century. It may be that Japan will find that is has generated enough plutonium for its ATR and breeder development programs from its present reprocessing arrangements and, thus, will delay the construction of a full-scale reprocessing plant.

Need for resolution of the spent fuel problem grows in Northeast Asia as well as around the world. The governments in all three countries realize they must provide temporary AFR storage and join in the study of both engineered permanent storage and geologic disposal. Engineered systems are especially attractive in all three countries because of seismic problems and lack of empty land. All three countries also have a strong long-term interest in the possibility of subseabed disposal of high-level wastes (or spent fuel). This option has support in Europe as well (Rippon, NN, 4/84, pp. 110–112) but is fraught with political and technical complexities (Hollister et al., 1981).

Each of the three countries is similarly uncertain about final disposal of low-level wastes. At present, each handles the wastes in somewhat similar ways: stored in drums, first at the nuclear plant and then in central facilities. In Korea, wastes are stored off-site. In China (Taipei), a small off-shore island, Lan-Yu (Orchid), has been used since 1982 and will be adequate for about 50 years. In 1987 the government, in response to growing public concern, announced that future low-level wastes would be sent to the United States (NN, 9/87, p. 90). Japan stores its low-level wastes, which in 1984 totaled more than 500,000 drums, in warehouses at centralized locations and has been considering land-based centralized storage schemes for three million drums (NN, 8/84, p. 77). A peninsula at the north of Honshu has been tentatively chosen as the site for the combined sites of low-level waste storage, reprocessing, enrichment, and centralized spent-fuel storage (NEI, 6/84, pp. 12–13). The AFR facility would initially hold about 100 reactor-years of spent fuel. When built, this facility would cost about US$4.1 billion, be operated by the Federation of Electric Power Companies (FEPC), and, most likely, would be the first such integrated fuel-cycle facility in the world (NEI, 9/84, p. 14).

Japan has also considered dumping its low-level wastes in the ocean. In spite of its intent to adhere to international agreements on such dumping (London Dumping Convention), Japan withdrew its plans to commence with a demonstration program because of diplomatic protests from a number of South Pacific nations in 1980 (Van Dyke et al., 1984).

NUCLEAR SAFETY

Although concern with safety is common to all nuclear utilities, it developed at different rates in different places as a consequence. For example, the Japanese were dramatically concerned over nuclear safety in the early 1960s when shortcomings were discovered in the country's first power reactor, the U.K. graphite-moderated Magnox. Problems surrounded the pressure vessel plate quality, graphite quality, and corrosion. This experience reinforced the natural Japanese view that quality assurance and reliability, features of Japanese industry on the whole, are particularly important in the nuclear sector. This view permeated the nuclear program from the start. It has also contributed to Japan's attitude toward international cooperation and independent development, leading Japan in part to explore many different reactor types, such as the KWU-PWR.

The implementation of safety measures in Japan involves complex interactions between the Atomic Energy Commission (AEC), the Ministry of International Trade and Industry, the Ministry of Science and Technology, the Nuclear Energy Bureau, the Nuclear Energy Safety Bureau, the Atomic Energy Safety Commission, and other groups. Consensus among many groups must be obtained, and two stages of public hearings are involved in the licensing and construction of any new plant.

In the 1970s frequent regulatory shutdowns and lengthy annual inspections were common occurrences and were reflected in low capacity factors. This was so in spite of the relatively small regulatory staff at MITI—about 50 in 1980.

Although there were no licensing requirements, most observers felt that Japanese reactor operators had better training than their American counterparts before the TMI accident. The regulatory apparatus was called into question in 1981 when it was revealed that contaminated water had been released from the 340 MW BWR Tsuruga plant south of Fukui. Along with allegations of a cover-up by the utility to avoid government and public scrutiny and potentially severe penalties, the resultant boycott of seafood from the region confirmed some of the worst fears of fishing communities about nuclear power plants. It seemed likely that this incident would exacerbate the already considerable local resistance to plant siting in many areas; it has made finding a shipyard for the nuclear-powered merchant ship *Mutsu* difficult.

In late 1978, the Nuclear Safety Commission (NSC) began operation. It was designed to start the process of separating regulation from promotion, functions that previously had been handled solely by the Japanese Atomic Energy Commission. The NSC oversees the work of MITI in regulation but does not implement safety regulations. It plays a crucial role in licensing procedure and has initiated local public hearings on reactor safety, held after MITI public hearings. The NSC hearings also cover a range of nonsafety topics.

Responding to the March 1979 TMI accident was the NSC's first real test. The accident precipitated inquiries from many local authorities, including pressure for shutdowns, and set the stage for a renewed series of antinuclear demonstrations around the country. The NSC responded by requiring MITI to undertake strict inspections of Japanese PWRs. At the time, only one PWR happened to be in operation. The others had been shut down for safety checks, but minor changes were made in the emergency systems of all PWRs. There were delays in granting permission to restart some PWRs, which resulted in low capacity factors, especially during 1979. Apart from a number of studies that were producing research, regulatory, licensing, design, and operational changes, concerns over TMI led to a gap in the granting of new construction licenses from the time of the incident until late 1980. Other outcomes included MITI's plans to install inspectors in every plant and to develop elaborate local emergency response plans. Finally, the NSC also issued reassurances to the Japanese public that a TMI-type accident could not occur in Japan because of design differences and more stringent regulations. The nine Japanese utilities have also been active participants in the analysis, research, and cleanup activities at TMI itself (NEI, 2/84, pp. 3–4).

In China (Taipei), nuclear safety is handled much as it used to be in the United States, at least in terms of organization, allowing that Taipower is government-owned. The Atomic Energy Council (AEC) has a Nuclear Regulatory Division, and one of its roles is nuclear reactor safety. Taipower also has a vice president that is concerned with nuclear safety as well as other areas of concern. The licensing procedure involves preliminary as well as final safety analysis reports patterned after the custom in the U.S. NRC. The peculiar seismic problems in China (Taipei) require special attention. For this purpose the quality assurance procedures developed in the United States were adopted.

In addition to the lowering of demand growth rate, an increasingly important reason for the continuing delay of Taipower's units 7 and 8 (station 4) is the rise in public concern over safety and waste disposal. This has been exacerbated somewhat by the USSR Chernobyl accident and by the July 1985 fire in one of Taipower's PWRs, which put it out of service for more than 15 months (Masters, 1987).

In Korea, nuclear safety research was established in the Korea Advanced Energy Research Institute (KAERI) in 1976. The safety regulations adopted combine elements from U.S. and Japanese regulations. KAERI uses Canadian standards for its CANDU reactor and U.S. standards for its Westinghouse PWRs. It has substantially upgraded its PWR safety-related equipment and reactor training since the mishap at TMI-2. Several hundred (300–600) technical people are needed each year, and a new nuclear engineering department has been established at the Ulsan Institute of Technology.

The prevailing view tends to be that reactor safety principally involves mechanical excellence, quality assurance of components, and inspections. But reports on the TMI-2 accident show that the people-machine interface is equally important, along with human perception of how events escalate in real time by means of mismatches between different parts of the system. In summary, the reactors might be built well, but the same reactor types may perform quite differently in different utilities.

During the first phase of nuclear capacity additions, Korea relied heavily on the regulatory programs of supplier countries (the United States, France, and Canada) to assure safety. Studies in recent years have recommended, however, that Korea develop a "strong, formalized, and self-sustaining Korean nuclear regulating program." Standards, review procedures, codes, and personnel would all require strengthening in quantity and quality. At present, ownership and management of the major institutions involved are intertwined, and the boundaries of responsibilities are confused. Perhaps more seriously, the Korean Atomic Energy Commission (KAEC), like the old USAEC, is responsible for both promotion and regulation, which are difficult objectives to pursue simultaneously. The KAEC was originally modeled after the organization in Japan, which has now split these roles into different agencies. But the 1983 International Atomic Energy Agency (IAEA) safety inspections in Korea seem to be the first of their kind.

Nearly all the nuclear reactor operations in the world outside the centrally planned economies associate with the Institute of Nuclear Power Operations (INPO) (NN, 10/84, pp. 67–85) and the Nuclear Safety Analysis Center (NSAC). INPO evaluates the effectiveness of nuclear power plant management programs, helps to upgrade the quality of operating personnel, and assists in the screening and evaluation of all licensing event reports for its member utilities. INPO also provides a computerized communications network called NOTEPAD to immediately inform all its member utilities of significant events at any plant and to facilitate the exchange of operational experiences of member utilities.

Taipower utilizes INPO's services directly, with stated satisfaction. All Japanese utilities participate in INPO through a representative company. There also may soon be a call for regional organizations such as an emergency preparedness team.

The TMI accident made a deep impression on utilities in Northeast Asia as it did elsewhere. In particular, the value of more systematic international cooperation was recognized because:

1. Nuclear power all over the world, or at least in all countries except certain centrally planned economies, is hostage to the least well run and most accident-prone reactors. Any large accident anywhere will adversely affect public and political acceptance of all reactors everywhere, particularly those of the same type.

2. Nuclear reactor system failures can have vast economic consequences for the electric utility system, even if the public hazard turns out to be negligible. Fireboxes, nozzles, blowers, superheaters, and other devices fail from time to time in fossil fuel plants, attracting no public attention at all. The plant is shut down, people enter any parts of it in a day or two, and repair small or large components at more-or-less standard industrial construction and repair costs. But accidents in a nuclear plant involve radiation hazards, may require extraordinary responses, and cost much more to repair or contain. The TMI and Chernobyl accidents are extreme examples.

The Chernobyl accident in 1986 (which occurred somewhat later than the period of primary interest in this book) has not apparently resulted in any institutionalized response within the utilities. This is due to the perception that technology and management conditions are so different in the Ukraine that useful information would not be available to Asian utilities. This is not to say, however, that utilities around the world will not be affected by the increased impetus of public criticism because of the Soviet accident.

FUEL SUPPLY AND SECURITY

It has been traditional to evaluate energy systems starting from the beginning of the fuel cycle, from mining and processing on to conversion, use, and waste disposal. Its placement here after discussions of reactor construction and operation, management, waste storage and disposal, and safety, however, is an indication that, for the foreseeable future, the uranium fuel cycle is not likely to be significantly constrained by supply of raw or processed fuel.

Japan has had a policy of diversifying all its sources of fuels, including uranium. Through 1990, for example, it has contracted to obtain a maximum of 35 percent of uranium from Canada. Although it has had contracts from Australia, South Africa, France, and the United States, unless new contracts are signed, after 1985 little will be coming from most of these areas. Japan will instead increase its reliance on Niger, the independent company Rio Tinto Zinc of London, and possibly Australia. There is little impetus for signing new contracts, because Japan has held stocks equivalent to more than a 10-year supply for its operating nuclear capacity since the early 1980s, and under current contracts and the most likely capacity growth (35 GW in 1990), it might hold as much as a 15-year supply in 1990. The economic cost of holding stocks is high. A 15-year supply might lead to additional electrical generation costs of 25 to 50 percent, depending on the carrying charges and the degree to which the uranium is held in enriched or fabricated forms (Neff and Jacoby, 1980).

The cost of this much energy security is high, especially since most observers believe that a buyer's market for uranium will persist throughout the decade and beyond. In 1984 uranium mines operated at only about 70 percent capacity, and less than 20 percent of the estimated global demand (outside of Communist countries) until the mid-1990s was uncommitted. In addition, in 1984 54 new uranium mines were in preparation which would in principle double existing world capacity, and many of the newer mines might have lower production costs than older mines. These factors plus the large stocks held in many countries mean that the uranium supply will not be a problem until the end of the century or beyond, barring, of course, political restrictions (Geddes, NEI, 6/84, pp. 15–21).

There is so much uranium held in stocks by consumer countries that significant amounts may continue to appear on the market for resale, as occurred during the early 1980s. Iran has been reselling uranium originally contracted from Namibia and South Africa as part of the former shah's ill-fated nuclear power program.

All three Northeast Asian countries have moved to acquire more control and experience in the upstream uranium market. Japan, for example, has an equity interest in Niger production, and Korea has been investing in the Gabon and U.S. uranium industries.

Although Korea has concluded small contracts with Canada and Australia, and China (Taipei) has a modest contract with South Africa, these two countries have not yet become major actors in the uranium market. This is largely because their nuclear programs are relatively recent, and initial cores and reloads traditionally are supplied by reactor vendors for new reactors. China (Taipei) has a strong policy of diversification in its fuel supplies, leading it to a range of buying practices including long-term contracts, spot purchases, loan agreements, purchasing reserves in place, and funding its own and other exploration and drilling projects.

Uranium, like petroleum, must be processed in special facilities before it can be used as an energy source. Relative to petroleum, the political and geographic distribution of present and potential exporters of significant amounts of uranium is large. In contrast to petroleum, processing capacity is restricted to a relatively small number of facilities. In 1980 this processing, called enrichment, was available for foreign use in significant amounts from only three suppliers—the U.S. Department of Energy (DOE) (nearly 70 percent of total commercial capacity), the Eurodif facility in France (20 percent), which is owned by France, Belgium, Spain, Italy, and Iran (which was selling its 10 percent share), and the Soviet Union, which has been offering enrichment services to non-Soviet bloc countries.

In the mid-1980s substantial excess enrichment capacity existed as a result of ever-optimistic estimates of uranium needs made in the 1970s, which stimulated several large construction programs. These were of three kinds: the classic, large gas diffusion plants, which are expensive and energy-intensive; centrifuges, which are much less energy-intensive; and laser isotope separation, which is also less energy-intensive and can substantially enrich the uranium in its fissionable U-235 isotope in one pass (compared to many passes in the other schemes), but is very technologically advanced. Two new European facilities are moving into operation—Coredif and Urenco (involving the United Kingdom, West Germany, and the Netherlands), one using gas diffusion and the other centrifuges. But at the same time the United States Department of Energy (USDOE), beset by low demand and high operating costs, planned in 1985 to close down its centrifuges and many of its gas diffusion plants, and to concentrate on perfecting laser isotope technology. Thus for the 1990s market shares of the various participants are hard to predict. Complicating the picture, Brazil and Japan planned to build commercial-scale plants for domestic use in this period as well.

Enrichment services, like uranium itself, will be available in abundance during the next decade. Present contracts will result in significant stockpiles of enriched uranium by 1990, at which time there will still be unused capacity. In 1980 the cumulative global surplus of enriched uranium equaled about a six years' forward supply, a surplus likely to remain until the end of the decade. By 1990 global demand will probably reach only 80 percent of enrichment capacity. Japan has bought nearly all of its enrichment services from the USDOE, even though a contract with Eurodif began in 1981. In Japan, demand for enrichment may be less than half the amount for which it has contracted by 1990. Indeed, Japan could well enter the 1990s with five to eight years of cumulative enrichment supplies. China (Taipei) and Korea have received all their enrichment services from the USDOE through contracts to purchase reactors from U.S. firms. Korea also has a 10-year contract with Eurodif as part of its 1980 deal to buy two French PWRs. For most of their other plants, Korea and China (Taipei) have 30-year enrichment contracts.

Japan constructed gas-centrifuge demonstration facilities at Ningyo-Toge for full operation by 1989 and has a commerical plant under construction at the new fuel-cycle site in north Honshu (starting at a capacity of servicing one and moving up to fifteen reactors). In light of the global enrichment excess, however, plans to add another equally large enrichment facility by the year 2000 can be attributed both to a heavy emphasis on energy security and as support for a Japanese nuclear export effort (NEI, 6/84, pp. 12–13; NEI, 12/87, pp. 46–47). It will be difficult, however, to compete in the open market, which will remain characterized by excess capacity in the large U.S. and European facilities even with the greater energy efficiency and flexibility promised by the centrifuge plants being planned in Japan.

Enrichment can be avoided entirely by employing reactors that use natural uranium directly. Heavy water reactors, the most popular of which is the CANDU, offer this advantage in addition to using about 15 percent less uranium and being somewhat less complicated to build. Their higher initial capital cost makes them less economic, however, unless uranium prices or insecurities rise significantly. Another approach that the Japanese are pursuing is to develop ATRs with higher fuel efficiency, requiring less enrichment effort per unit of energy output.

ECONOMICS AND TRADE

The nuclear industry in Northeast Asia is important not only in terms of total investment and GNP, but also in terms of balance of trade, which to date has been mainly with the United States. The finances and financing of large electric power projects are complicated subjects and cannot be handled in detail here (see, however, Chapter 6). In this section we touch briefly on the major components of cost and the implications of the changing character of the export market on Northeast Asia.

How Much Does Nuclear Power Cost?

Nuclear power plants are expensive and show surprising cost variability: In the United States, for example, superficially similar plants in the early 1980s cost from about US$1,000/kW to US$4,000/kW. This variability has led to confusion in many minds. The nuclear part of a nuclear-electric power plant costs, or should cost, only about 15 percent (at the most 20 percent) of the total plant cost. All the coal handling, boiler, and effluent treatment equipment of a coal-fired plant do not cost much less. Some of the increased cost of the nuclear plants arises from the lower steam temperatures that are produced (compared to fossil fuel plants), hence the need for larger and less efficient components. But even so, one must look elsewhere for the often considerable cost premium of nuclear power plants.

What usually happens is that a combination of siting, regulatory, A/E, scheduling, public acceptance, and managerial problems arising because of nuclear-specific issues causes the nuclear part to apply much economic leverage to the entire plant. These problems slow its construction and lead to rework and backfits. As evidence, some nuclear power plants are built in the United States on time, within budget, with few unforeseen regulatory problems, and are of good quality. In all three Northeast Asia countries, the construction record is relatively much better. Fewer alternatives to nuclear power surely influence attitudes toward both acceptance and construction times. The relatively high cost of coal in Northeast Asia (US$60–70/ton delivered) leads to a coal fuel cost alone of 2.2–2.6¢/kWh, comparable to US$1,000/kW capitalization with 14 percent return on investment for a 70 percent capacity factor. Oil-fired power plant fuel costs would be about twice as high.

It is surprisingly and frustratingly difficult to compare the costs of electric power generation even in different parts of one country (Spinrad, 1980). In comparing countries the difficulties are compounded because of currency exchange distortions, extreme variations in accounting practices, and entirely different financial arrangements (not to mention differences in the date, type, size, and operating history of the plants and the varying amounts of infrastructure that must be built along with them) (NN, 3/84, pp. 88–89). In addition, such comparisons are difficult when, as in Korea and China (Taipei), significant portions of the financing come through export-import bank loans with low-risk premiums or subsidized rates (Hayes et al., 1984).

When comparisons are made, however, it seems that nuclear power retains an advantage over coal for large baseload plants in all OECD countries except the United States (NEA, 1984, 1986). The ratio is somewhat smaller in Northeast Asia, but it does seem generally to favor nuclear over coal. A 1984 study by the Korean Ministry of Energy and Resources, although finding nuclear cheaper than coal, identified potential future increases in nuclear capital and waste-disposal costs. This, combined with the much lower amounts of foreign exchange required for coal plants, means that the advantage to nuclear may be narrow at present (NEI, 1/84, p. 9). Furthermore, by early 1987 in Japan, the price of oil in yen had decreased to the extent that nuclear plants were more expensive than oil or coal once estimated waste disposal and decommissioning costs were included (NN, 3/87, p. 95).

To say that nuclear-electric power has a potential cost advantage is not the same as realizing it. In order to overcome some of the difficulties noted above, certain cost-reduction measures should be undertaken by all countries in the future. Both exogenous and endogenous factors affect these costs. Regarding exogenous factors, interest charges and price escalation on nuclear power plant construction and on nuclear fuel procurement are very influential in all three countries, because the power plants are not largely self-financed. Instead, the fraction of external funds is as high as about 0.6 in Japan, and 0.5 in Korea and China (Taipei). Thus the installer of a nuclear power plant needs to take effective measures to shorten the lead time, especially the construction period.

Another factor in this period that has not been critical in Asia but is prominent in the United States is the perceived riskiness of nuclear investments among both utilities and investors. It is mainly financial risks rather than direct safety risks that have combined with lower electric demand to stall U.S. nuclear power plant orders and construction projects (NEI, 9/84, p. 13). It is estimated, for example, that nuclear utilities paid a risk premium of about 2 percent for investment capital during the early 1980s (EW, 7/84, p. 9). This can make a big difference in a capital-intensive industry. To date there have been no financial disasters in Asia such as TMI or the U.S. bond defaults. Although they have not been completely immune to events in North America, this particular affliction of nuclear power has yet to be transported across the Pacific.

Studies have shown that financial risk is dominated by replacement power costs during short-term outages. In addition, essentially all accident scenarios result in higher on-site financial losses rather than off-site ones. To some extent, the increasing rate of forced outages has been caused by regulatory actions. In addition, operation and maintenance (O&M) costs in U.S. nuclear plants increased dramatically (almost by a factor of four) from 1975–1982, whereas O&M costs at coal plants increased by less than 50 percent (NEI, 6/84, p. 10).

Regarding indigenous factors, price escalation of construction costs requires keen attention. Innovative research and development to reduce costs is necessary, as is emphasis on standardization to reduce costs on repeat items. Additionally, efforts should be made to accurately identify and reduce back-end fuel cycle costs, even though they may appear small in present cost estimates. Such costs should also include decommissioning.

Recent high interest rates have, of course, aggravated the problem of nuclear plant costs. High interest rates tend to lead to short time perspectives. Although many analyses show that on a long time scale for financial return, large nuclear power plants would be cheaper than the alternatives, the money market does not work that way. Lack of short-term liquidity, in the words of the financier, works against nuclear and other long-term options. Indeed, it has been argued that the upward uncertainty of oil prices tends to increase the effective interest and discount rates by which capital-intensive energy alternatives are evaluated, ironically leading to greater attractiveness of oil use (Smith, 1982).

Nuclear Trade Futures

No other country except Japan has developed such a comprehensive nuclear industrial capability before starting to export (Lester, 1983). For a number of years such a move has seemed inevitable, although there are some barriers. It would seem, however, that to compete effectively with the existing supplier nations, such an export business would have to be supported by a much-expanded Japanese fuel cycle industry. Some observers think that Japanese industry is not ready for exports and should cooperate directly with its traditional American partners—Westinghouse and General Electric—in any initial export plans. Indeed, in late 1984 Westinghouse and Mitsubishi announced a joint $1.5 billion bid for construction of a Turkish PWR (NN, 10/84, p. 81). The 1986 annual report of the Japanese Atomic Energy Commission, however, indicates that Japan may well decide to proceed alone (NEI, 2/87, pp. 21–22).

Japan's Asian neighbors would be natural customers for a Japanese export market, and a few steps have been made in that direction. In mid-1984, Mitsubishi announced a contract to supply the 300 MW reactor vessel for China (Beijing)'s first power reactor. In the same period, several Japanese banks agreed to lend to the Korean nuclear power program (NN, 7/84, p. 62). In addition, a joint nuclear construction company has been established between Taipower and Mitsubishi (QER, 1981, p. 48). These developments could well be the start of a combined effort on the part of the Japanese industrial and banking communities to cooperate on nuclear bids.

The Westinghouse partnership with Mitsubishi may be the trend of the future in U.S. nuclear trade because of a decrease in construction capacity accompanying a lapse in domestic orders (Lester, 1984). This is likely to result in a gradual diminution in the international technological and market influence of U.S. firms. As Richard Lester has pointed out, this may lead to greater dominance of European exporters in Asia but may also result in a gap that can be capably filled by the Japanese. To do this Japan would have to take on regional political roles quite different from those it has developed since the Second World War.

A major change could occur with regards to Japanese prospects for exports should the Japanese be able to develop a commercially viable small power reactor, something long discussed in the world industry (Murota, 1983) in recent years along with the related concept of an intrinsically safe reactor (Wakabayashi, 1985).

SOCIETAL ISSUES

Large commitments to nuclear power generate societal issues. Among these are the trade-offs between the social and economic costs and benefits of more electric power; public acceptance or rejection, whether overt or covert; resource independence vis-à-vis interdependence; and questions of intergroup equity. Two of the more basic issues—siting and international security—are briefly discussed below.

Siting

Siting has major social implications. In all three countries there seems to be relatively little support for the idea that nuclear power is totally unnecessary. Thus the debate tends to focus on where and when to install it. Relative to Europe, North America, and the Philippines, not much public opposition exists at present, even in Japan. Seacoast siting is virtually mandatory in all three countries.

Japan's famous nuclear allergy may be older, but it does not seem to be as strong as the allergies in some other OECD countries (Tanaka, 1984). One consequence, however, is that locating suitable and acceptable sites is extremely difficult, a problem exacerbated by Japanese geography. As Table 5.4 indicates, about half of Japan's power plants are owned by two utilities, Kansai and TEPCO. This has decided advantages for quality assurance over the U.S. situation in which many small utilities have one reactor but do not have a large technical backup staff or extensive experience in nuclear plant operations. Indeed, TEPCO is the largest privately owned utility in the world and has considerable economic and political influence in Japan as well as technical expertise. This concentration of power and technical ability means that Japan has been moving de facto into what has been termed "nuclear parks." These are emerging at the research headquarters—Tokai at Fukushima south of Sendai, and in the region around Wakasa-wan Bay on the Sea of Japan. At Shimokita at the north end of Honshu, two utilities are planning to develop jointly a 20,000 MW site with the first reactor expected to be on-line in 1990. To help alleviate the siting problem and to respond to seismic safety concerns, the Japanese government has been conducting feasibility studies to determine the viability of underground and marine siting.

The reliability of this option is uncertain, but the emergence of more autonomous local decision making has led to a substantial reduction in available coastal sites. It has also led to what are essentially transfer payments from urban to rural dwellers in the form of subsidies to local areas that do accept power plants. For a large nuclear power station, typically 3.1 billion yen went to the nearby community and another 3.1 billion yen to neighboring communities during a period that varied between five and ten years. This amounts to about 2 percent of the construction cost, which may increase the cost of delivered nuclear electricity by 1 percent.

Enough good sites exist in all three countries for all planned reactors. In Korea, four major sites could accept a total of 20 units up to the year 2000. Residents and fishermen are compensated for land taken or the exclusion of fishing in nearby waters. To forestall any future antinuclear activities, KEPCO carries on a public education program (e.g., teams are sent to give presentations and show slides and films).

In Korea, the scheduled total lead time to build nuclear plants is 12 years; usually the plan calls for awarding the contract after 2.5 years of planning and beginning excavation after 4.5 years. These lead times are also affected by the availability of suitable sites, and the 44 plants once planned for the year 2000 would require nearly 20 sites. Several factors, however, limit the number of suitable sites: seawater is generally used for cooling; coastal fishing industries are important; many coastal areas belong to the military; and all sites must be located below 37 degrees latitude for security in the event of a north-south conflict. Indeed, the Republic of Korea has said that it has foiled at least two commando raids from the north directed against its Wolsung CANDU (NN, 10/83, p. 75).

In China (Taipei), 18 units could be put on 4 sites, but beyond that, siting would be more difficult. Furthermore, China (Taipei) believes the complete filling of present sites might cause some difficulties. Among these is the idea that risk should be assessed collectively at a site with more than one unit and the total risk for a multiunit site should be limited, not just the risk per reactor. Public acceptance, favorable until recently, appeared less sure in 1984–1985. Taipower realizes the necessity of maintaining good public opinion and has programs such as seminars and movies shown throughout the island, with special attention given to groups near existing sites. It also cooperates with government agencies to conduct extensive education programs for high school teachers.

Siting also remains a problem because of the high seismicity of most of the island. Four sites at the northern and southern tips of the island have been chosen for the first eight reactors, and design modifications have been incorporated into imported U.S. equipment because of this problem. In general, however, China (Taipei) has relied on U.S. safety criteria and regulatory approaches. It has also been successful in expediting its reactor purchases by requiring that U.S. vendors receive government export approval before submitting bids. In 1981 this policy was approved by the U.S. Nuclear Regulatory Commission (USNRC) along with an agreement not to allow intervendor hearings during the export license procedure for China (Taipei).

International Security

Nuclear power has at times appeared in an ambiguous role, stimulating a two-sided debate about its stabilizing and destabilizing effects. It can reduce tension by reducing anxiety and insecurity related to lack of access to reliable energy resources, but it can also increase tension because of the fear that the manufacture of weapons will proliferate and become easier. In fact, the civilian nuclear programs of some countries have been interpreted as nothing more than a cover for a weapons program. In the earliest periods of their nuclear industries, these types of charges were leveled at Korea and China (Taipei).

Civilian nuclear power plants also make prime targets in warfare, especially if sited along seacoasts. A country that installs nuclear power reactors becomes a potential hostage and must maintain relatively peaceful behavior so as not to destabilize the region by appearing to subvert the civilian power system for weapons use. This aspect, relevant to all three countries, is often overlooked.

This vulnerability expresses itself in different ways in the three countries. Japan appears not to take any extraordinary security measures against sneak attack. China (Taipei) takes some measures, and Korea takes considerably more, although unspecified. Although such measures might well protect the plants from ground (or sea) attacks on a modest scale, they cannot prevent the plant from being rendered inoperative by a well-targeted conventionally armed missile. In this sense, a significant fraction of the nation's power supply is hostage to peace.

The Non-Proliferation Treaty (NPT) is a cornerstone of international security with respect to nuclear power. It was envisaged as an effort to stop both "vertical" and "horizontal" proliferation. The NPT committed nonnuclear weapons signatories to refrain from acquiring such weapons (Article II) and to accepting safeguards (Article III). It committed signatories with nuclear power programs to assist in the legitimate tasks of developing nuclear power around the world (Article IV), and it committed the nuclear weapons signatories to halting and de-escalating their nuclear weapons races (Article VI).

Surely some ambiguity arises in interpreting Articles I and II vis-à-vis Article IV, because civilian nuclear technology and expertise can be co-opted to assist in weapons programs (although there is more to this than just building and delivering bombs). The imposition of more stringent controls on civilian nuclear trade by the nuclear supplier nations, coupled with the near total failure of the United States and Soviet Union to pay the slightest attention to Article VI, has been a source of serious frustration for many developing countries. Were the NPT to be proposed de novo today, many developing nations would probably not sign it, out of irritation over the commitments on their part and little or none by the nuclear weapons states.

Despite these difficulties, the three Northeast Asian countries support the NPT and other such activities. Japan is a member of the NPT, accepts IAEA international safeguards, holds bilateral agreements with five countries (the United States, Canada, the United Kingdom, Australia, and France), and participates in the London (supplier) guideline negotiations. It engages in active cooperation to improve the technology and development of international safeguards. Plutonium is to be recycled to the ATR, followed by the LWR, and perhaps eventually to a commercial breeder after 2010.

Korea is a signatory to the NPT (December 1975) and also has bilateral agreements with the United States, France, Canada, and Australia. It has actively participated in a range of international dialogues related to nuclear power (e.g., the International Fuel Cycle Evaluation, or INFCE, the Committee on Assurance and Supply, or CAS, and the International Plutonium Storage, or IPS). As discussed earlier, Korea stores spent fuel either on-site for 10 years or at AFR storage for 30 years. If reprocessing is permitted, Korea hopes to utilize a Pu-thermal reactor followed by a breeder around 2010–2020.

China (Taipei) was dismissed from the IAEA when China (Beijing) was admitted. Nevertheless, China (Taipei) continues to support the NPT fully and hopes for the joint formation of an international safeguards package for civilian nuclear power programs that would be amenable to identical interpretation by both suppliers and consumers. Technical experts in China (Taipei) look toward reprocessing for better waste management, which would eventually be followed by plutonium utilization in the breeder. It now keeps spent fuel stored either on-site for 10 years or at AFR for 20 years.

CONCLUSIONS

The likely future development of nuclear power in Northeast Asia depends on more than the topics discussed here. The following stand out:

1. Japan has a well-developed nuclear industry, appears to be in or near a position of technical leadership in LWR technology, and has excess reactor manufacturing capacity. With the introduction of additional fuel cycle capabilities, it could compete effectively for international sales.

2. Japan is prepared to build 300–500 MW LWRs as well as larger ones and is seriously engaged in research and development on advanced reactors. These developments will strengthen its international marketing ability.
3. Korea and China (Taipei) have large nuclear commitments, although the rate of expansion is now declining. Both have localization programs.
4. Korea's burgeoning manufacturing and A/E capability in many areas besides nuclear makes it a serious competitor internationally, especially in heavy industry.
5. Although not discussed here, China (Beijing) is committing itself to nuclear by simultaneously developing its own reactor as well as purchasing two French PWRs (NEI, 9/87, p. 26). If the history of the region is any guide, it may well develop an impressive domestic capability before the end of the century.
6. The degree to which Korea and China (Taipei) continue to rely so heavily on nuclear power may depend on the path each takes to full political democratization. As has been the case in the Philippines and elsewhere in the Asia-Pacific region (Van Dyke et al., 1984), nuclear issues often become intertwined with political struggles in a way that acts to magnify public concerns over safety and waste disposal.

The U.S. energy role in the Asia-Pacific region is liable to decline, except for specific items. Those exceptions are unlikely to include nuclear for several reasons. First, the U.S. nuclear sector does not appear to be healthy, despite recent efforts to stimulate it to new activities. Second, the United States has not lived down its reputation since the late 1970s as an unreliable partner. Third, although U.S. vendors and A/E firms have business in East Asia at present, Japan, Korea, and China (Taipei) are expanding such activities. And finally, the U.S. nuclear waste program appears unlikely to be very helpful to Asia-Pacific nations, except for some general science and technology.

The movement of China (Beijing) to the center stage of Asia-Pacific energy affairs becomes more certain each day. The potential for nuclear imports, the impact on nuclear demand of coal exports, and the potential for nuclear waste disposal in the deserts of China (Beijing) will influence the future developments of nuclear power in Asia.

Japan may well emerge as the leader over the United States, dealing with China (Taipei) and Korea as advanced junior partners and sometimes as competitors, with China (Beijing) as a major trading partner, and perhaps with the ASEAN nations as customers over the longer term.

Advantages emerge in regional training and supply centers, and as professional personnel become trained predominately in the region, their intellectual connection with other regions will decline (Masters, 1987). In this highly technological arena, trade patterns will thus be substantially affected.

Many Asia-Pacific nations, observing what happened successively in Japan, China (Taipei) and Korea, may opt for rapid technological growth, including energy technologies and (selectively) nuclear power. Safety analyses and the operating experiences in countries worldwide (the majority relatively event-free), the post-TMI advances in LWR technology, the somewhat strengthened international arrangements (e.g., INPO), the modest-to-moderate-but-apparently-continuing economic advantage of nuclear electricity (except near large coal reserves), and the rising concern over the environmental problems of coal all combine to make nuclear power somewhat more attractive and less uncertain than was the case a few years ago. It remains to be seen, however, whether other developing countries can master the management, financial, regulatory, and public participation efforts required to take advantage of nuclear power's potential.

NOTES

1. Although this history ends before the Chernobyl accident in 1986, few of the major trends identified here have changed (see, e.g., Masters, NEI, 6/87, pp. 25–27).
2. PWRs and BWRs are the major types of LWRs.
3. Canadian deuterium (heavy water) reactor.
4. As of early 1987, at 72 percent there was essentially no difference in the average load factors (NEI, 7/87).

REFERENCES

Chu, David S. L. (1984). "Nuclear Power Program in Taiwan: Present and Future." *Energy—The International Journal,* 9(9/10):779–788.

Electrical World (EI). Issue and page references in text.

Hayes, P., Shorrock, T., and Zarsky, L. (1984). "Korea, Nuclear Bonanza." *Multinational Monitor* 5:17.

Hirata, Mitsuko (1984). "Institutional Arrangements for Nuclear Fuel Cycle Services." *Energy—The International Journal* 9(9/10):883–888.

Hollister, C. D., Anderson, D. R., and Heath, G. R. (1981). "Subseabed Disposal of Nuclear Wastes?" *Science* 213:1321–1326.

Industrial Bank of Japan (1982). "Japan's Industry at a Turning Point." *Japan's Finance and Industry,* Quarterly Survey No. 51, (July–September).

Lester, Richard K. (1982). "Nuclear Power Plant Lead-times." In Ian Smart, ed., *World Nuclear Energy.* Baltimore: Johns Hopkins University Press.

——— (1983). "Structural Change" in the "Nuclear Power Plant Industry in the Pacific Basin Region." *Columbia Journal of World Business* (Fall) 18:16–24.

——— (1984). "The Nuclear Power Industry in the Asia-Pacific Region." *Energy—The International Journal,* 9(9/10):875–882.

Masters, R. (1986). "World Survey: Still Working Through the Backlog." *Nuclear Engineering International* 31(383):46–79.

______ (1987). "Asian Pacific Basin: Finding a Catalyst for Cooperation." *Nuclear Engineering International* 32(398):39–45.

Murata, Hiroshi (1984). "Nuclear Power Development in Japan." *Energy—The International Journal* 9(9/10):751–760.

Murota, Kenji (1983). "Status of Development of Small Reactors in Japan." Paper presented at the conference on Nuclear-Electric Power in the Asia-Pacific Region, Resource Systems Institute, East-West Center, Honolulu (January).

NEA (1984). "The Costs of Generating Electricity in Nuclear and Coal-fired Power Stations," Nuclear Energy Agency, Organisation for Economic Co-operation and Development, Paris.

______ (1986). "Projected Costs of Generating Electricity from Nuclear and Coal-fired Power Stations for Commissioning in 1995," Nuclear Energy Agency, Organisation for Economic Co-operation and Development, Paris.

Neff, Thomas, and Jacoby, Henry (1980). *The International Uranium Market* MIT-EL-80-014. Cambridge, MA: Energy Lab, MIT.

Nuclear Engineering International (NEI). Issue and page references in text.

Nuclear News (NN). Issue and page references in text.

Nucleonics Week (NW). Issue and page references in text.

Payne, J. (1987). "South Korea: Planning for Self-reliance." *Nuclear News* 30(15):59–63.

Quarterly Economic Review (QER). Issue and page references in text.

Smith, Kirk R. (1982). "Nuclear Power in Asia-Pacific Region." In F. Fesharaki et al., *Critical Energy Issues in Asia and the Pacific.* Boulder, CO: Westview Press.

Spinrad, B. I. (1980). Effects of Accounting Rules on Utility Choices of Energy Technologies in the United States, RR-80-27, International Institute for Applied Systems Analysis, Laxenburg, Austria.

Sung, N. C. (1983). "The Nuclear Industry in Korea." Paper presented at the Conference on Nuclear-Electric Power in the Asia-Pacific Region, Resource Systems Institute, East-West Center, Honolulu (January).

Tanaka, Yasumasa (1984). "Nuclear Issues in the Asia-Pacific Region: Socio-political Factors." *Energy—The International Journal,* 9(9/10):909–918.

Van Dyke, J., Smith, K. R., and Siwatibau, S. (1984). "Nuclear Activities and the Pacific Islanders." *Energy—The International Journal,* 9(9/10):733-750.

Wakabayashi, H. (1985). "Forgivingly safe and safeguarded small and intermediate sized reactors everywhere in the world." Paper presented at the second workshop on Nuclear Electric Power in the Asia-Pacific Region, Korea Power Engineering Co., Seoul.

CHAPTER 6

FINANCIAL MANAGEMENT AND CORPORATE ORGANIZATION

Y. H. Kim

This chapter looks closely at the financial and tariff patterns of the Japanese utilities, the Korea Electric Power Corp. (KEPCO), and Taipower. Because expertise inside the electric enterprises is mobilized through different management structures, the chapter ends with comparisons of utility management organizations in Northeast Asia.

MANAGEMENT EFFICIENCY

The electricity supply industry is capital intensive and requires extensive funds for year-by-year investment in additional generation, transmission, and distribution facilities. Particularly during periods of high inflation, high interest rates, and low load growth, such as the late 1970s and early 1980s, financial considerations have been a prime concern in electric utility decisions and plans.

Financial Soundness

As the indicators in Table 6.1 show, the financial positions of these utilities in Northeast Asia deteriorated in 1973. In particular, KEPCO reached a critical stage that year, with a negative internal contribution to investments, a rate of return on investments that dropped to about 4 percent, a debt-service coverage ratio of only 67 percent, and a ratio of total liabilities to net worth that drastically increased. The main reasons for this unfavorable trend were temporary overinvestment in generation facilities, insufficient tariff adjustments for achieving a reasonable proportion of internal financing, and, as a consequence, excessive borrowing at comparatively short-term, high interest rates. Since then, the financial performances of KEPCO and the Japanese

Table 6.1
Financial Indicators and Planning Methods, 1970–1980

	Japanese Utilities	KEPCO	Taipower
1. Financial indicators			
Rate of return (%)			
1970	8.4	5.6	12.6
1973	4.8	4.2	11.1
1980	11.8	11.5	11.6
Self-financing ratio (%)			
1970	36.0	20.0	77.0
1973	14.0	-17.0	34.0
1980	37.0	11.0	30.0
Debt-service ratio (%)			
1970	240.0	130.0	173.0
1973	105.0	67.0	143.0
1980	181.0	117.0	82.0
Debt/equity ratio (%)			
1970	268.0	176.0	88.0
1973	387.0	331.0	131.0
1980	614.0	258.0	153.0
Average annual rate of increase in real price of electricity, 1970–81 (%)	5.5[a]	3.7	5.2
Average annual rate of increase in real cost of electricity, 1972–81 (%)	6.2[b]	5.3	7.3
2. Depreciation practice			
Straight-line method	Yes (1973)	Yes (27—>18 yr)	Yes (26—>15 yr)
Fixed rate method	Yes (1980)	No	No
3. Financial targets			
Rate of return	At least 8%	At least 9%	At least 9.5–12.0%[c]
Self-financing ratio	NA	At least 20%	At least 30–40%[c]
Debt-service ratio	NA	At least 125%	At least 150%[c]
4. Financial planning method			
Utility financial model	Yes	Yes	Yes
Energy sector integration	Yes	No	No
Fiscal and monetary sector	Yes	No	No

Sources: Ke et al. (1983); Kim et al. (1983); Kimura et al. (1983).

a. 1970–1980. c. By Taipower calculation method.

b. 1972–1980. NA Not applicable.

utilities have improved greatly (see item 1 in Table 6.1). Taipower has maintained a somewhat better financial position over the entire period.

The financial health of utility companies is partly determined by the rates they are allowed to charge for electricity. The price of electricity in the three countries increased substantially from 1970 to 1981, even when measured in real terms, but in general these increases have merely offset the rapidly rising real costs of the electric utilities (see item 1 in Table 6.1). Because the average annual rate of increase in the real cost of electricity between 1970 and 1980 substantially exceeded that of its real price by 0.7 percentage points in the Japanese utilities, 1.6 percentage points in KEPCO, and 2.1 percentage points in Taipower, the three systems were able to maintain their sound financial situations only at the cost of increasing their debt burden. As a consequence, the ratio of total liability to net worth increased substantially in all three countries between 1970 and 1980.

The future in the three countries seems likely to bring significant financial difficulties, caused chiefly by the excess baseload capacity that resulted from slower load growth and a lower load factor than in the past, higher interest rates, and a higher proportion of capital-intensive units in new generating facilities. In the long run, however, a higher proportion of nuclear generation will substantially reduce fuel costs, thereby improving the financial situation in the three systems. Meanwhile, all three will have to endure a severe financial bind during the next five years and possibly longer. Several financial and management rationalization measures have been put into effect by the Japanese utilities, KEPCO, and Taipower, and important price adjustments are to be implemented.

The extent of internal financing is principally determined by the rate at which capital assets can be written off against revenues, since depreciation provisions are by far the most important source of internal funds. Important changes in accounting procedures have been made recently in all three countries. In an attempt to alleviate the financial burden on utility authorities, which resulted from lower rates of increase in tariff rates, Japan adopted the straight-line method of depreciation in 1973. After the second oil crisis, when the tariff rate increased by 45 percent in 1980, the Japanese utilities adopted fixed-rate depreciation. Most fixed assets, however, are currently depreciated on a straight-line basis. The extent to which the fixed-rate method is applied to fixed assets will be increased in steps, whenever the effects on the revised tariff of adopting the fixed-rate method become excessive.

The Japanese utilities are also reviewing the introduction of inflationary or current-price depreciation in an effort to increase depreciation allowances. In contrast, both KEPCO and Taipower are employing straight-line depreciation. A recent change in accounting procedures shortened the assumed lives of fixed assets from 27 to 18 years in KEPCO and from 26 to 21 years in Taipower, thereby increasing the annual depreciation charges by substantial amounts. The major differences in accounting procedures among the three countries lie in the handling of capital revaluation. Whereas the Japanese utilities value their capital stock at its installation cost, both KEPCO and Taipower revalue theirs periodically in the calculation of depreciation. Consequently the Japanese capital stock, unlike that of its two neighbors, will be undervalued throughout its lifetime because of inflation.

In all three countries financial models are generally used in conjunction with capacity expansion models (see item 4 in Table 6.1). Utility financial models are used in the three countries to provide information on capital requirements, cash flows, and rates of return. In the case of Japan the utility financial model for the electric power sector is integrated with the energy sector model as well as the fiscal and monetary models of the Japanese economy. The financial targets are somewhat different in the three countries (see item 3 in Table 6.1).

Comparative Prices

As electric supply authorities in every country aim to recover their accounting expenditures, changes in the major components of average cost per unit of electricity sold can be evaluated to determine the principal reasons for differentials in the price of electricity that existed in the three countries.

Table 6.2 compares average unit prices for residential, commercial, and industrial sales among the three countries. Average price figures in current prices were deflated by the respective price indexes of each country to allow for general inflation. The resulting figures expressed in 1975 constant prices were then converted to U.S. dollars as a common unit using 1975 official exchange rates. Although this procedure ignores the effects of purchasing power differences among the three countries during the period 1970 to 1981, it nonetheless appears to be justified because the principal reason for making these international comparisons is to trace changes that occurred within each country.

Between 1970 and 1981 average prices paid by household consumers in the three countries remained more or less at constant levels in real terms with some minor fluctuations between years (Table 6.2). In Korea the average residential price increased at almost precisely the same pace as general inflation in 1975, whereas there was a fall in real prices in both Japan and China (Taipei). This reflects Korea's firm determination to conserve energy immediately after the oil crisis in 1973–1974. In contrast, the intentions of its two neighbors were to protect household consumers against high inflation in 1974–1975. On the other hand, residential consumers in Korea enjoyed a price advantage over their counterparts in China (Taipei) and Japan during the whole period under review. There is a general postulation that increasing electricity prices at a rate higher than inflation stimulates conservation; if aver-

Table 6.2
Average Unit Prices for Electricity Sales in Real Terms, 1970–1981 (1975 US cents/kWh)

	1970	1975	1980	1981	Index (1970 = 100) 1975	1980	1981
Residential sales							
Japanese utilities	6.88	5.27	6.97	7.05	77	101	102
KEPCO	6.55	6.40	5.46	5.59	98	83	85
Taipower	3.77	2.53	3.56	4.56	67	94	121
Commercial sales							
Japanese utilities	4.26	5.05	7.50	7.57	119	176	178
KEPCO	3.18	5.54	9.48	10.00	174	298	314
Taipower	5.75	5.14	6.80	8.03	89	118	140
Industrial sales							
Japanese utilities	2.45	3.24	4.90	4.95	132	200	202
KEPCO	2.12	2.79	3.94	4.03	132	186	190
Taipower	1.90	2.63	3.27	3.82	138	172	201
Total sales							
Japanese utilities	3.35	3.91	5.68	5.77	117	170	172
KEPCO	3.12	3.53	4.67	4.82	113	150	155
Taipower	2.55	2.81	3.58	4.42	110	140	173

Sources: Ke et al. (1983); Kim et al. (1983); Kimura et al. (1983).

age electricity prices were growing at approximately the same rate as inflation, the result could actually be an increase in electricity use, because other energy sources such as oil and gas would be increasing at rates higher than inflation during the same period. This observation, however, should be modified depending on the extent to which different tariff rates for each income bracket increased or decreased relative to general inflation.

Korean commercial consumers in 1970 enjoyed a great advantage over those of China (Taipei) and Japan, as Table 6.2 shows, but this relative price advantage disappeared after the oil crisis in 1973–1974. More dramatically, the real price of electricity paid by commercial users in Korea increased by more than three times between 1970 and 1981. This represents a conscious government effort to save electricity for nonproductive users as part of a comprehensive conservation program. As a consequence, in Korea the proportion of electricity sales to commercial consumers declined considerably from 20 percent in 1970 to 13 percent in 1980.

It is apparent from Table 6.2 that the average industrial price in real terms in China (Taipei) remained the lowest among the three countries throughout the period. Although the industrial price differential between China (Taipei) and Korea narrowed in 1981, the gap between China (Taipei) and Japan widened persistently. Because electricity is a major element in the production cost of electricity-intensive materials, and in light of Japan's higher prices, China (Taipei) and Korea might gain a comparative advantage over Japan in future international trade in the electricity-intensive materials that these countries produce. China (Taipei) and Korea are in competition, however, not just with Japan but with all other industrialized countries, and it appears that a more detailed study has to be made of potential market size, potential scale, and other economies in the use of electricity in the more heavily electricity-dependent production processes. In this regard, it is remarkable how rapidly the Japanese industrial structure has already undergone a change from an electricity-intensive stage to a technology-intensive path of industrialization. Despite stagnation in the production of electricity-intensive materials—for example, iron, steel, aluminum, cement, petrochemicals, pulp, and paper—the production of technology-intensive materials such as machinery and electronics has increased dramatically. This rapid structural change in Japanese industry can be attributed partly to a relatively higher price of electricity in real terms. Likewise, as a result of sharp electricity price increases after the second oil crisis in 1979–1980, the comparative advantages in China (Taipei) and Korea have shifted against electricity-intensive process industries.

Industrial consumers in the three countries of Northeast Asia paid less per unit of electricity than either residential or commercial users throughout the period, as Table 6.2 indicates. On the other hand, commercial users in 1981 were paying more (approximately 175 to 250 percent in KEPCO and Taipower) per unit than either industrial or domestic users, the differential being the largest in Korea. Thus both KEPCO and Taipower are penalizing their commercial as well as residential users for conservation reasons. Japan takes a different approach, charging real costs to the different types of service as a function of voltage characteristics and their impact on distribution facilities.

In all three countries the overall average prices of electricity increased at a higher rate than general inflation in the period 1970 to 1981. It appears that this real price increase has stimulated conservation. During the same period China (Taipei) maintained the lowest overall average price in real terms among the three countries; Japan had the highest. By 1981 Korea managed to reduce, although only slightly, the considerable price advantage held by China (Taipei).

Tariff Structures

As described above, the financial positions of the Japanese utilities, KEPCO, and Taipower have improved significantly since 1973. From 1973 to 1981 important adjustments to average rates or to specific rates for some categories of customers were made in the wake of the sharp cost increases, as evidenced by the comparison of trends in average price and average cost per kWh. Some of the adjustments in the three countries were aimed at increasing the amount of internal contributions to investment. As a consequence, between 1973 and 1980 the self-financing ratio increased greatly from –17 to 11 percent in KEPCO and from 14 to 37 percent in the Japanese utilities (see Table 6.1). Other rate readjustments were made to reduce the subsidies to some classes of customers in order to establish a more equitable rate structure.

Past rate increases, particularly in Korea, have been aimed at reducing inequalities in the rate structure. A comparison of supply costs and average revenues by customer category is shown in Table 6.3. Although the agricultural sector in Korea and the residential sector in China (Taipei) are still subsidized, the cost-of-service data indicate serious efforts by KEPCO and Taipower to establish a more adequate rate structure so that the price of electricity sold to each customer will be closer to the cost of providing the service. The commercial sector, on the other hand, is heavily penalized. This above-cost pricing of electricity to commercial users reflects both governments' firm determination to foster energy conservation.

The most compelling theoretical framework for pricing is on the basis of long-run marginal costs (LRMC) (see Chapter 3 and Munasinghe and Warford, 1982). LRMC pricing will eventually lead the utility system to its least-cost mix of reliability and supply sources, assuming some stability in external factors. In practice, however, power systems are sometimes so far from the optimum that LRMC pricing is not possible. In this case, short-run mar-

Table 6.3
Cost of Service by Major Customers, 1978–1981 (ave. revenue/supply cost)

	Residential Sales	Commercial Sales	Industrial Sales	Agricultural Sales	Street Lighting	Total
KEPCO						
1978	1.11	1.71	1.15	0.37	1.01	1.09
1980	1.03	1.77	1.11	0.46	1.05	1.11
1981	1.07	1.83	1.04	0.45	1.08	1.08
Taipower						
1978	0.74	1.71	1.05	—	0.68	1.02
1980	0.73	1.64	0.95	—	0.80	0.94
1981	0.81	1.79	1.00	—	0.93	1.03

Sources: Ke et al. (1983); Kim et al. (1983); Kimura et al. (1983).

— Not available.

ginal cost (SRMC) pricing has the advantage of drawing the demand into a pattern that utilizes the existing system most effectively. For example, Taipower has been using SRMC pricing because of its excess capacity. Figure 6.1 shows the SRMC for 200 MW of extra power both on- and off-peak as estimated by Taipower in 1985–1986. Other considerations must also be included in final tariffs, of course, including rate of return, rate of loss for each customer class, and so on.

In all three countries, some incentives for conserving energy have appeared in the tariff structure since the first oil crisis of 1973–1974 (see Chapter 3). The residential and commercial consumers in China (Taipei) and Korea—but only residential consumers in Japan—are on an increasing block schedule, so that as consumers use more electricity the overall price per kWh increases (see items 1 and 2 in Table 6.4). This is in sharp contrast to the United States where most consumers are on either declining block schedules or flat rates. The effect of increasing block schedules on energy saving turned out

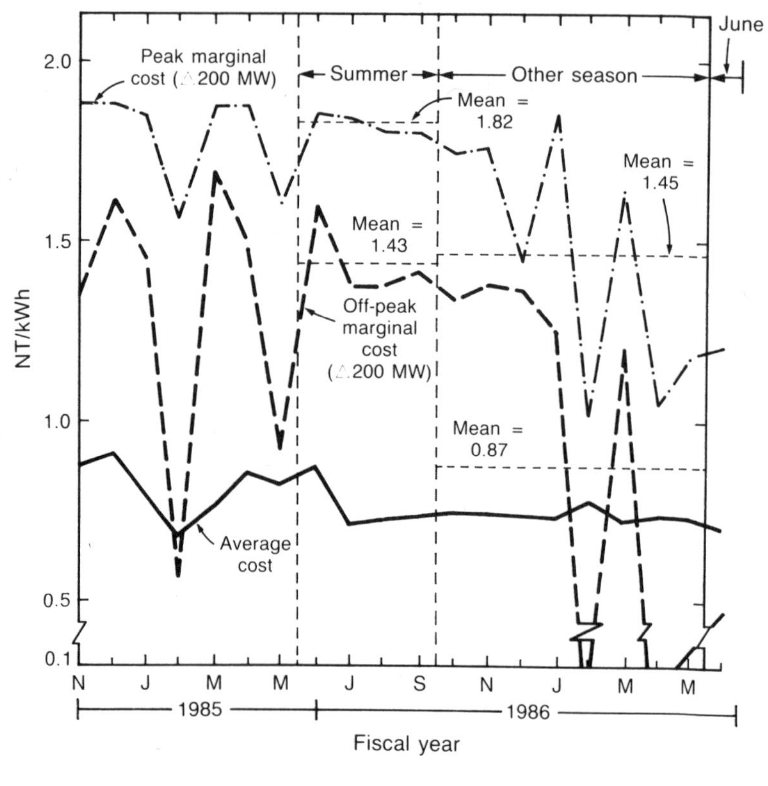

Figure 6.1. Average and short-run marginal costs for Taipower by month.

Table 6.4
International Comparison of Tariff Rate Structure

	Japanese Utilities	KEPCO	Taipower
Residential services	1. Two parts 2. Demand charge (proportional to contracted kW or ampere) Energy charge (three blocks increasing rate; 1:1.6)	1. Two parts with minimum charge 2. Demand charge (proportional to kWh used) Energy charge (six blocks increasing rate; 1:11)	1. Single part 2. Energy charge (three blocks increasing rate; 1:1.7)
Commercial services	1. Two parts 2. Demand charge (flat rate per contracted kW) Energy charge (flat rate per kWh used) 3. Seasonal rate for energy charge (10% higher for summer use)	1. Two parts 2. Demand charge (flat rate per contracted kW) Energy charge (five blocks increasing rate; 1:1.7 for public use; 1:3.4 for commercial use)	1. Two parts 2. Demand charge (flat rate per contracted kW) Energy charge (three blocks increasing rate; 1:1.3)
Industrial services	1. Two parts 2. Demand charge (flat rate) Energy charge (flat rate with 10% summer seasonal adjustment) 3. Special contract for interruptible power (applicable to extra-high voltage customers)	1. Two parts 2. Demand charge (flat rate) Energy charge (flat rate) 3. Time-of-use rate (energy charge for contracted kW exceeding 300 kW) Peak : partial peak : off-peak = 3:1.5:1	1. Two parts 2. Demand charge (flat rate) Energy charge (flat rate) 3. Time-of-use rate (energy charge for contracted kW exceeding 100 kW) Peak : off-peak = 1.7:1

Sources: Ke et al. (1983); Kim et al. (1983); Kimura et al. (1983).

to be enormous. In China (Taipei) the implementation of increasing block schedules resulted in a sharp decline in the average annual growth rate of household sales per customer from 5.4 percent in 1977–1978 to 2.5 percent in 1980–1981. It should be noted that Taipower is subsidizing its residential customers as a whole for the purpose of consumer protection (the "ability to pay" principle), and that Taipower managed to reduce the growth rate of household electricity consumption by applying increasing block rates. Commercial sales per contracted kW also declined, but to a lesser degree, from 177 kWh in 1977–1978 to 171 kWh in 1980–1981.

Of equal importance to structural changes in residential and commercial rates, Japanese utilities, KEPCO, and Taipower replaced the previously decreasing rate schedule for industrial customers with a flat rate applicable to all energy consumption. Perhaps the most significant adjustment to the pricing of electricity is that, for load management purposes, KEPCO and Taipower introduced peak-load pricing for large industrial customers (see item 3 in Table 6.4). KEPCO instituted time-of-use rates in December 1977 for industrial customers whose contracts exceed 300 kW. Figure 3.14 illustrates the results in some major industries. In August 1979 Taipower likewise began applying time-of-use rates sparingly to industrial customers that contracted more than 100 kW, with the provision that industrial users contracting between 100 kW and 500 kW could opt for time-of-use rates. This followed studies made on long-term marginal cost pricing for electricity. Whereas peak energy costs for the largest industrial customers are three times higher than off-peak rates in KEPCO, the ratio of peak rates to off-peak rates is 1.7 in Taipower. In Japan the special contract rate system for interruptible power was designed for load management purposes (see Chapter 3).

Early indications show that such a pricing policy is effective in shifting the power demand of large industrial customers to off-peak hours. Some studies in other countries indicate that reductions in system load on the order of 1

Table 6.5
Effect of Time-of-Use Rates in KEPCO, 1976–1980

	Load Profiles		
	1976 (before time-of-use rate)	1979	1980
Off-peak load	80	83	84
Partial peak load	101	100	104
Peak load	127	121	115
Average load	100	100	100

Sources: Ke et al. (1983); Kim et al. (1983); Kimura et al. (1983).

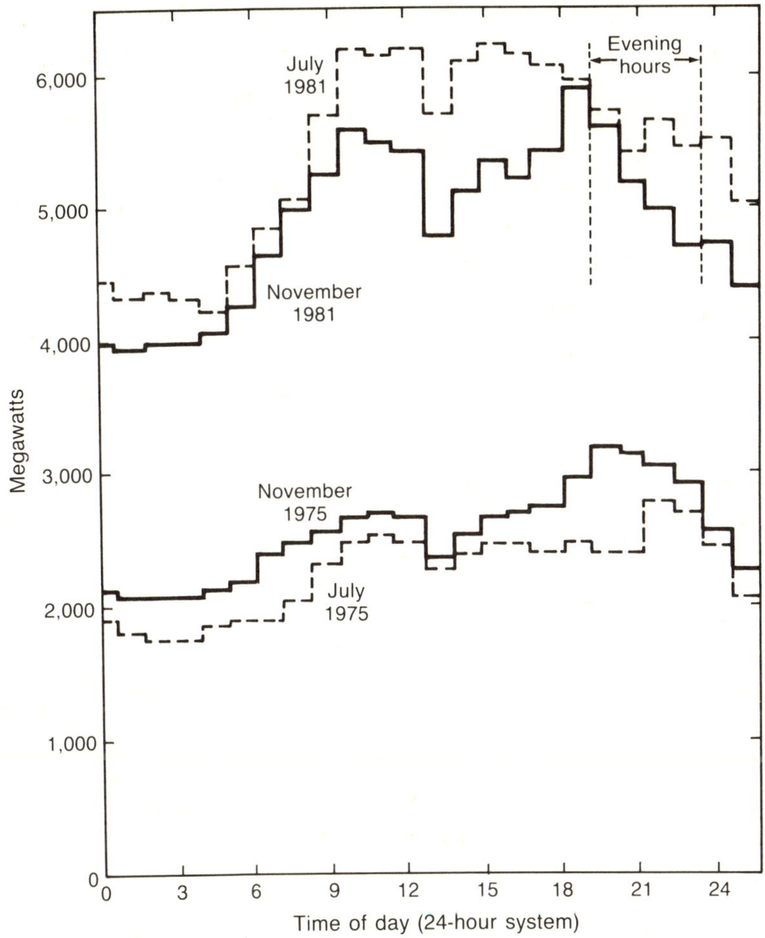

Figure 6.2. Daily load curves for July and November in 1975 and 1981 for KEPCO.

to 10 percent are possible, applying time-of-use rates. KEPCO experienced an approximate 10 percent reduction in peak load between 1976 and 1980 (see Table 6.5). As a result the load curve for large customers now shows a significant dip during the evening hours as industries respond to the strong price incentives (see Figure 6.2). This illustrates the effectiveness of marginal cost pricing and load management techniques. Nonetheless, a recent KEPCO survey revealed that major industrial and commercial customers are using their own emergency backup diesel generators to offset their industrial demand during peak hours and to avoid the cost impact of the increasing block structure for commercial use. It should also be noted that KEPCO's load curve shifted from the former winter evening peak to the present summer morning and afternoon peaks (10–12 a.m. and 2–5 p.m.). Neither of the double peaks actually occurred within the time range of the peak tariff (7–10 p.m.). Accordingly, KEPCO is expected to adjust the peak tariff period to reflect the time of actual peak loads.

In summary, the many rate adjustments in Japanese utilities, KEPCO, and Taipower that have taken place since the 1973–1974 oil crisis have considerably modified their tariff structures toward more efficient execution. The concept of conservation in these countries was introduced to reduce the absolute amount of electricity consumption after the first oil shock. After the 1979–1980 oil shock, there was a shift in thinking to more "rational and effective" use of electricity generating facilities. This cannot be separated from the load management need to maximize the overall capital efficiency of the economy. In considering the possible financial difficulties resulting from excess baseload capacities that are likely to exist in the three countries until at least the late 1980s, restructuring the tariff rate system to adapt the load characteristics to those of the generation mix will be important in conjunction with possible construction schedule delays.

As part of sales promotions to create new baseload demand, Japan is introducing heat pumps for cooling and heating and has already implemented hot- and chilled-water storage techniques. The latter are also under consideration in China (Taipei), together with the introduction of more effective time-of-use rates and special tariffs for interruptible power. To further pursue the principle of long-term marginal cost pricing in Korea, the peak-load pricing system can be extended to smaller industrial customers and others. In general, one of the most critical management issues is how to reduce generating costs until the low-cost nuclear era is realized, in view of the future prospects of lower load growth, prominent summer peak, and built-in excess baseload capacity.

Looking back on the past 20 years, the electric power industries in all three countries smoothly and successfully carried out both management rationalization and—as national goals—interfuel substitution, first away from coal and later away from oil. Management rationalization was realized during a period of rapid economic growth, high load growth, and (up to 1973) cheap fuel prices. Looking forward to the coming 20 years, the utility managements in all three countries, to varying degrees, are now faced with the harsh reality of carrying interfuel substitution even further away from oil, while at the same time maintaining the efficiency of the power industry. The future prospects include lower economic growth, lower load growth (characterized by prominent summer peaks but stagnant baseload), and high capital intensities of nonpetroleum-based technologies.

International Comparisons of Electricity Costs

Despite the earlier comparison of electricity prices, it is difficult to easily draw conclusions concerning the relative management efficiency of the three utility systems. The objective of private companies exposed to market competition is to maximize profits; management efficiency is therefore often measured in terms of profits. In contrast to competitive private companies, the profits of electric enterprises are regulated by the government. As public utilities, electric enterprises, whether privately or government owned, aim to minimize the electricity supply cost while securing an adequate supply capability. The fair-return principle is commonly adopted by the electricity enterprises as their pricing rule. In making an international comparison of the management efficiency in the electric power industry, therefore, it is more appropriate to compare production costs than to compare average electricity prices.

The data in Table 6.6 represent the principal cost factors contributing to increases in the real price of electricity in China (Taipei), Japan, and Korea from 1970 to 1981. There is no compelling economic reason why costs should be similar in three countries that cannot engage in international trade in electricity. Costs may differ because resource availability affects the costs of fuel, capital, and labor. In comparing costs between the three countries, therefore, special attention should be given to the costs of fuel and capital.

In all three countries there was a significant rise in the average cost of fuel used per unit of electricity sold when measured in real terms (see item 1 in Table 6.6). The extent of increases in fuel cost and its level in real terms varied: from 1970 to 1981 fuel costs in real terms were lowest in China (Taipei) and highest in Korea. Moreover, fuel cost differentials widened in these countries over time. In general, fuel cost differences were attributable to changes in the prices of fuel delivered to power stations, in operating efficiency of the stations, and in generation mix.

To meet a higher demand growth, the power sector's investment was substantially increased during the period 1960 to 1983. In keeping with important technological progress in electricity production associated with larger equipment, all three countries significantly increased the size of generating equipment installed. In the case of KEPCO, for example, the average size of thermal power plants sharply increased from 22 MW in 1960 to 209 MW in 1983, thereby taking advantage of economies of scale, particularly fuel economies (see item 5 in Table 6.7). As item 4 of Table 6.7 indicates, between 1960 and 1983 the thermal efficiency considerably improved—by 5.8 percent in Japanese utilities, 9.0 percent in Taipower, and 12.0 percent in KEPCO. This higher thermal efficiency reduced the amount of physical fuel required to produce a unit of electricity. It should be noted that the gaps in thermal efficiency between countries narrowed in 1983. Since the electric power supply industries in China (Taipei) and Korea heavily relied on imported supplies of new generating equipment, it appears that no long delays occurred in taking advantage of technological advances made in Japan. There is considerable evidence that the electricity authorities in the three countries had substantially narrowed the technological gap by 1983.

In addition to fuel economies achieved by more efficient use of fuel in new generating stations, fuel consumption was also reduced through sharp decreases in transmission and distribution losses (see item 3 in Table 6.7). Part

Table 6.6
Average Unit Costs of Electricity by Major Components in Real Terms, 1970–1981

	1970		1975		1981		Rate
Component	Cost[a]	(%)[b]	Cost[a]	(%)[b]	Cost[a]	(%)[b]	(%)[c]
1. Fuel cost							
Japanese utilities	0.57	17.7	1.48	37.8	2.17	39.0	12.9
KEPCO	0.74	29.2	1.94	56.5	2.61	58.5	18.6
Taipower	0.72	35.9	1.33	53.8	1.94	52.3	9.4
2. Salaries, wages							
Japanese utilities	0.58	17.9	0.56	14.3	0.48	8.6	1.7
KEPCO	0.52	20.5	0.28	8.2	0.22	4.9	7.5
Taipower	0.19	9.4	0.20	8.1	0.25	6.7	2.5
3. Other operating costs							
Japanese utilities	1.15	35.6	1.09	27.9	1.70	30.6	3.6
KEPCO	0.44	17.3	0.62	18.1	0.81	18.2	5.7
Taipower	0.42	20.9	0.36	14.6	0.50	13.5	1.6
4. Interest charges							
Japanese utilities	0.36	11.1	0.47	11.9	0.58	10.4	4.4
KEPCO	0.43	16.9	0.30	8.7	0.48	10.8	1.0
Taipower	0.29	14.4	0.22	8.9	0.39	10.5	2.7
5. Depreciation charges							
Japanese utilities	0.57	17.7	0.32	8.1	0.63	11.4	0.9
KEPCO	0.41	16.1	0.29	8.5	0.34	7.6	1.7
Taipower	0.39	19.4	0.36	14.6	0.63	17.0	4.5
6. Total							
Japanese utilities	3.23	100.0	3.92	100.0	5.57	100.0	5.1
KEPCO	2.54	100.0	3.43	100.0	4.46	100.0	5.3
Taipower	2.01	100.0	2.47	100.0	3.71	100.0	5.8

Sources: Ke et al. (1983); Kim et al. (1983); Kimura et al. (1983).

a. Constant 1975 US cents per kWh.

b. Component cost per kWh as a percentage of total electricity cost.

c. Average annual growth rate 1970–1981.

Table 6.7
International Comparison of Technical Efficiency Indicators, 1960–1983

Indicator	1960[a]	1965[a]	1970[a]	1975	1980	1983
1. Load factor (%)						
Japanese utilities	67.4	68.7	69.8	60.7	62.8	60.6
KEPCO	67.1	61.6	67.3	67.6	77.7	73.4
Taipower	65.0	69.1	70.8	69.4	69.3	66.6
2. Demand density (MWh/km^2)						
Japanese utilities	36	76	140	220	281	329
KEPCO	12	25	78	169	331	431
Taipower	101	158	334	591	1,058	1,180
3. Rates of transmission and distribution losses (%)						
Japanese utilities	11.3	8.5	6.8	6.4	5.8	6.0
KEPCO	29.4	19.2	11.8	11.3	6.7	6.3
Taipower	13.6	12.1	9.1	7.1	6.8	6.8
4. Thermal efficiency (%)						
Japanese utilities	29.8	34.7	35.9	36.4	36.4	36.6
KEPCO	22.6	24.0	29.7	31.7	33.6	34.9
Taipower	26.5	31.4	34.4	35.0	35.8	35.5
5. Average unit size (MW)						
Japanese utilities	110	239	316	443	456	481
KEPCO	22	32	68	131	165	209
Taipower	20	50	100	199	246	321
6. Fossil thermal generation share (%)						
Japanese utilities	51.0	61.4	77.1	74.8	67.1	66.1
KEPCO	63.2	78.1	86.7	91.1	84.7	76.4
Taipower	42.6	65.5	80.8	79.2	74.3	49.8
7. Plant factor (%)						
Japanese utilities	59.4	53.9	60.5	46.4	44.2	43.9
KEPCO	52.6	48.3	41.7	48.0	45.1	42.5
Taipower	50.3	55.2	61.6	48.0	51.1	41.3
8. Labor productivity (MWh/employee)						
Japanese utilities	649	893	1,583	2,057	2,522	2,715
KEPCO	157	234	528	948	1,720	2,157
Taipower	441	565	1,002	1,441	2,106	1,919

Sources: Ke et al. (1983); Kim et al. (1983); Kimura et al. (1983).

a. 1961, 1966, and 1971, respectively, for Taipower.

of this success in China (Taipei) and Korea was due to implementation of measures to sharply curtail power pilferage. This is a difficult management issue in most developing countries, and one with sensitive political overtones in some.

An even more important factor in the decrease in fuel consumption was the substantial increases in nuclear generation, which reduced the share of fossil thermal generation in total generated power (see item 6 in Table 6.7). On the other hand, between 1972 and 1981 the unit price of fuel oil in real terms increased greatly in all three countries, with the average annual rate of increase ranging from 15 percent in China (Taipei) to 17 percent in Korea and 18 percent in Japan. This substantial increase so dominated the effects of the favorable factors contributing to the reduction of fuel consumption that the average price of fuel measured in real terms increased persistently in all three countries between 1970 and 1980 (see item 1 in Table 6.6). Whereas the rate of increase in the real price of fuel oil among the three countries was fastest in Japan, Korea had the highest unit cost of fuel and the highest rate of increase when measured in real terms. This was because Korea's total power generation also had the lowest proportion of fuel-saving units (nuclear and hydropower). It is also apparent from items 3, 4, and 5 in Table 6.7 that Korea achieved the fastest technological progress in electricity production among the three countries over the period 1960–1983, thereby considerably narrowing the technological gaps.

Whereas China (Taipei) had the lowest relative salaries and growth rates in real terms between 1970 and 1981, both Japan and Korea achieved savings in salaries measured in real terms during the same period (see item 2 in Table 6.6). In China (Taipei), on the other hand, salaries per kWh continued to increase during the same period, particularly in 1981.

More interesting is the comparison of labor productivity as item 8 in Table 6.7 shows. In all three utility systems, the net sales of electricity per employee rose significantly during the period 1960–1983, thereby considerably reducing the unit cost of labor. Whereas the differences in labor productivity reflect different factor costs in each country, the labor productivity gap between the three countries narrowed considerably in 1983. In this respect, it is remarkable how rapidly KEPCO raised its labor productivity in the past 23 years to close the gap, first with Taipower and later with the Japanese utilities.

As the capital-cost data in Table 6.6 show, interest charges in real terms in Korea were more or less constant from 1970 to 1980, whereas both Japan and China (Taipei) experienced a substantial increase. Depreciation charges, when measured in real terms for 1970–1980, declined slightly in both Japan and Korea. There was, however, a drastic increase in China (Taipei) in 1981 (see item 5 in Table 6.6). The load factor of KEPCO substantially improved from 67 percent in 1970 to 73 percent in 1980 (item 1 in Table 6.7), whereas in both Taipower and the Japanese utilities the load factors continued to deteriorate between 1970 and 1980. On the other hand, Taipower had the lowest interest and depreciation charges among the three countries. This is so because among the three utility systems demand density and plant factor were highest for Taipower (see items 2 and 7 in Table 6.7), which, for Taipower, resulted in the highest utilization of capital stock.

ORGANIZATION

The three countries of Northeast Asia have different types of organizations for the control and development of electric power. This section covers the industry structure, labor force, and management structure of the three.

Industry Structure

In China (Taipei) the sole enterprise for power development and supply is the Taiwan Power Company (Taipower), a government-owned undertaking with limited private ownership of less than 6 percent. The company operates a power network that covers practically the entire island. Taipower was incorporated under the National Enterprise Law of the Republic of China in 1946, after having taken over the power facilities on Taiwan and Penghu from the former Japanese company. Under the provisions of the National Enterprise Law, the Ministry of Economic Affairs regulates and supervises Taipower's activities but also shares legislatively mandated powers on matters of personnel, accounting and finance, purchasing, and business with the Executive Yuan, Ministry of Audit Affairs, and the Legislative Yuan (the parliament).

In Korea there are three operating companies. The Korea Electric Power Corporation (KEPCO), fully owned by the government, is the sole distributor of electric power throughout the entire country. It owns and operates the bulk of the country's total generating capacity. Except for multipurpose projects, KEPCO is responsible for the planning and construction of all power projects. The Kyongin Energy Company is a private generating company with an installed capacity of 325 MW whose entire output is sold to KEPCO. The Industrial Site and Water Development Company (ISWACO), a government-owned entity, owns and operates multipurpose projects with a total installed capacity of 380 MW and sells its power output to KEPCO. The Korea Electric Power Company (KECO) was established in July 1961 through a merger involving the Chosun Electric Power Generation and Transmission Company, the Kyungsung Electric Power Distribution Company, and the Namsun Electric Power Distribution Company, under the Korea Electric Power Company Act. The KECO Act was amended in 1976 to provide for full government ownership by the end of 1981. KECO was then renamed the Korea Electric Power Corporation.

Several Korean government agencies participate in regulating electric power. The Ministry of Energy and Resources controls KEPCO's activities under the Electric Enterprise Law (1973). It also shares the approval of KEPCO's budget with the Economic Planning Board and the Ministry of Finance, the control and development of nuclear power with the Ministry of Science and Technology, and the establishment of tariff rates with the Economic Planning Board. On the other hand, the Ministry of Construction builds, owns, and operates all multipurpose projects directly or through ISWACO.

Table 6.8
Electric Power Utilities in Japan, March 1984

Companies	Supply Area (km^2)	Number of Customers (10^3)	Generation	
			Hydro	Thermal
General electric power companies				
Hokkaido Electric Power Company	78,522	2,808	1,026	3,079
Tohoku Electric Power Company	79,546	5,728	2,243	6,255
Tokyo Electric Power Company	39,463	19,009	5,058	23,676
Chubu Electric Power Company	39,131	7,473	3,483	13,952
Hokuriku Electric Power Company	12,282	1,501	1,693	1,912
Kansai Electric Power Company	28,643	9,779	5,859	12,991
Chugoku Electric Power Company	32,134	4,175	1,650	5,561
Shikoku Electric Power Company	18,452	2,227	774	3,171
Kyushu Electric Power Company	42,129	6,141	1,626	7,377
Subtotal of 9 companies	370,302	58,841	23,412	77,974
Okinawa Electric Power Company	2,249	501	0	976
Total	372,551	59,342	23,412	78,949
Wholesalers				
EPDC	NA	NA	6,489	3,493
JAPCO	NA	NA	0	0
Others	NA	NA	2,498	11,139
Total electric utilities	372,551	59,342	32,400	93,580
Industry-owned	NA	NA	1,126	13,685
Grand total	372,551	59,342	33,525	107,265

Source: TEPCO.

NA Not applicable.

— Not available.

Private enterprises hold a dominant position in the electric power industry of Japan. The country has been divided into nine monopolized service territories, with a privately owned company operating in each under a license or franchise. The nine regional electric power companies (Hokkaido, Tohoku, Tokyo, Chubu, Hokuriku, Kansai, Chugoku, Shikoku, and Kyushu) operating today were inaugurated in May 1951 (Table 6.8 and Figure 6.3). In addition to these companies, the Okinawa Electric Power Company supplies electric power to the Okinawa area. The Electric Power Development Com-

Table 6.8 (*continued*)

Capacity (MW)		Energy Produced (10^6 kWh)	Consumption (10^6 kWh)			Peak Demand (MW)
Nuclear	Total		Lighting	Power	Total	
0	4,105	16,820	5,337	11,510	16,847	2,990
0	8,498	35,882	10,787	29,255	40,042	6,990
6,896	35,630	143,348	39,912	109,378	149,290	32,380
1,380	18,816	71,880	14,513	55,923	70,436	14,235
0	3,605	12,164	3,173	13,550	16,723	3,135
5,668	24,518	92,793	22,563	68,345	90,908	19,958
460	7,671	25,037	7,994	26,443	34,427	6,824
1,132	5,077	18,987	4,326	12,503	16,829	3,391
1,118	10,120	37,543	12,309	29,653	41,962	9,139
16,654	118,040	454,454	120,914	356,550	477,464	98,886
0	976	3,849	1,303	2,058	3,361	—
16,654	119,015	458,303	122,217	358,608	480,825	NA
0	9,982	33,662	0	0	0	NA
1,623	1,623	10,135	0	0	0	NA
0	13,637	53,394	0	18,964	18,964	NA
18,277	144,257	555,493	122,217	377,572	499,790	NA
165	14,975	62,607	0	53,261	53,261	NA
18,442	159,232	618,100	122,217	430,833	553,050	NA

Note: Peak demand is the top three daily peak averages of the month of August (except Hokkaido, which is based on February).

pany (EPDC) was jointly established in 1952 by the government and the nine regional power companies to undertake the construction and operation of large-scale, remote, and expensive hydroelectric projects and coal-fired thermal power generation projects. Although it is a government-controlled entity (the government owns 72 percent of EPDC's stock), it sells its entire output

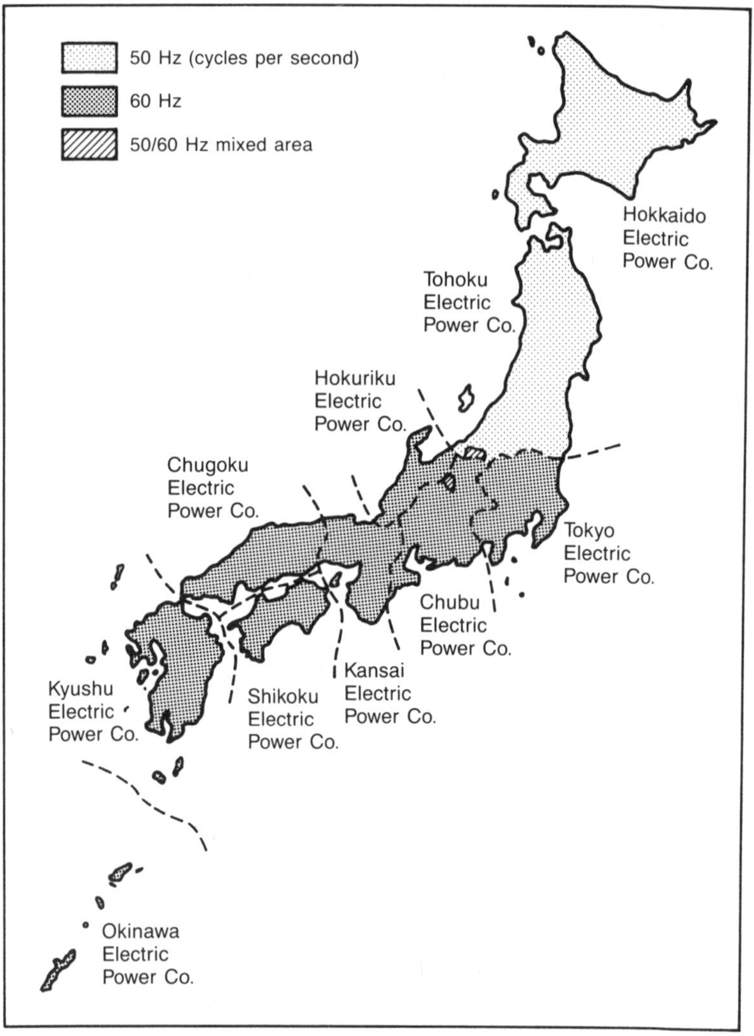

Figure 6.3. Service areas of Japanese electric utilities. Note that line frequency is different in northern and southern Japan.

to the nine electric power companies. More deeply related to these nine companies is the Japan Atomic Power Company (JAPCO), established in 1957 and engaged in the country's nuclear power development projects from the initial stage. The nine electric power companies own 75 percent of JAPCO's stock, and JAPCO wholesales power to the nine companies. There are also 54 minor undertakings that sell power wholesale to the nine companies, which control approximately 74 percent of the total installed capacity of the industry.

Whereas the Japanese electric supply industry is wholly in the hands of private enterprise, the electric power companies are regulated and supervised by the Ministry of International Trade and Industry (MITI) under the provisions of the Electric Enterprise Law (1964) and other legislation. MITI, however, shares with the Science and Technology Agency and the Environmental Agency legislatively mandated powers to approve the construction of new plants, to establish rates, and to regulate safety and certain environmental standards. MITI and its advisory Electric Power Development Coordination Council must approve construction and siting of new plants. MITI can also exercise some influence over decisions of future power plant mix by having EPDC build some plants and by providing subsidies to alleviate siting problems. But the electric power companies will provide most of the incremental nuclear and coal-fired power generation in the next decade. Accordingly, implementation rests largely in the hands of the nine electric power companies.

What most distinguishes the Japanese electric utilities from KEPCO and Taipower are private ownership, relative management autonomy, and freedom from government control. Both KEPCO and Taipower rely heavily on government sources for the massive capital required for the expansion of power production and transmission facilities. Raising the necessary capital from the open market is relatively easy for the Japanese electric utilities. Accordingly, these utilities have ample room for autonomous action in taking initiatives and making investment decisions on future power development. The Japanese electric utilities are more important within Japan than are KEPCO in Korea and Taipower in China (Taipei). The nine privately owned regional electric power companies, led by the Tokyo Electric Power Company, exert significant influence over Japan's power development policies and strategies. Their economic power stems more directly from their enormous capital budget and purchasing power. It can also be argued that this has the reverse impact as well, and that it is not possible for the Japanese electricity supply industry to move rapidly from the present power plant mix to the nationally desired mix of power plants.

Whereas government control over utility management is desirable from the point of view of national economic growth, government intervention inhibits the exercise of ingenuity and managerial creativity of electric utilities. This is recognized in Korea where, in an attempt to provide greater management autonomy to KEPCO, the Korean government promulgated the Government-Owned Corporation Management Law on March 1, 1984.

Labor Force

Unlike Japanese utilities under private management, both KEPCO and Taipower are government owned and vested with the authority to supply electricity throughout the whole country and to exploit and develop power resources. For several decades, both companies have made strenuous efforts and attained reputations for success in supplying electricity to keep pace with the growing demand. One of the key factors in this achievement is their resourceful personnel management.

By 1981 KEPCO had 19,600 employees; Taipower had 24,700. Personnel employment by functional category for the period 1961–1981 for KEPCO and 1950–1981 for Taipower is shown in Tables 6.9 and 6.10, respectively. Two elements in the tables are worth observing. One is that employment growth rates (3.4 percent for KEPCO during 1961–1981 and 5.2 percent for Taipower during 1950–1981) are much lower in comparison to those of generation, indicating an improvement in labor productivity. Another is the gradual shift (more

Table 6.9
KEPCO Employment by Functional Categories, 1961–1981

Year	Administrators	Engineers	Technicians	Total
1961	3,116	5,775	1,204	10,095
	(30.9)[a]	(57.2)	(11.9)	(100.0)
1965	3,209	5,793	2,979	11,981
	(26.7)	(48.4)	(24.9)	(100.0)
1970	2,985	5,106	4,349	12,440
	(24.0)	(41.0)	(35.0)	(100.0)
1975	3,012	6,084	4,863	13,959
	(21.6)	(43.6)	(34.8)	(100.0)
1980	3,896	9,354	5,276	18,526
	(21.0)	(50.5)	(28.5)	(100.0)
1981	3,992	10,012	5,575	19,579
	(20.4)	(51.1)	(28.5)	(100.0)
Average growth rate (%)				
1961–65	0.7	0.1	25.4	4.4
1965–70	0.2	-2.5	7.9	0.8
1970–75	0.2	3.6	2.2	2.3
1975–81	4.8	8.7	2.3	5.8
1961–81	1.2	2.8	8.0	3.4

Source: Kim et al. (1983).

a. Numbers in parentheses are percentages of the annual total.

prominent for KEPCO) of percentage ratios among function categories, which means operational safety and reliability have been given priority since increasingly sophisticated elements have come into the system.

KEPCO, however, has encountered problems in retaining employees over the long term. As Table 6.11 indicates, employees with employment periods shorter than five years accounted for nearly 45 percent of all employees in KEPCO and fewer than 30 percent in Taipower (compared with fewer than 20 percent in Electricite de France). Those with service periods exceeding 20 years are 4.3 and 23 percent in KEPCO and Taipower, respectively (compared with 33 percent in EDF). The phenomenon is more pronounced for KEPCO engineers, many of whom, although college graduates, were less than 30 years

Table 6.10
Taipower Employment by Functional Categories, 1950–1981

Year	Administrators	Engineers	Technicians	Total
1950	1,603	278	3,291	5,172
	(31.0)[a]	(5.4)	(63.6)	(100.0)
1955	1,620	607	4,197	6,424
	(25.2)	(9.5)	(65.3)	(100.0)
1960	2,140	1,496	5,147	8,783
	(24.4)	(17.0)	(58.6)	(100.0)
1965	3,196	1,704	6,224	11,124
	(28.7)	(15.3)	(56.0)	(100.0)
1970	3,507	2,502	7,895	13,904
	(25.2)	(18.0)	(56.8)	(100.0)
1975	3,330	4,347	10,518	18,195
	(18.3)	(23.9)	(57.8)	(100.0)
1980	5,170	6,098	11,545	22,813
	(22.7)	(26.7)	(50.6)	(100.0)
1981	5,547	6,396	12,753	24,696
	(22.5)	(25.9)	(51.6)	(100.0)
Average growth rate (%)				
1950–60	2.9	18.3	4.6	5.4
1950–70	4.0	11.6	4.5	5.1
1950–81	4.1	10.6	4.5	5.2
1960–70	5.1	5.3	4.4	4.7
1960–81	4.6	7.2	4.4	5.0
1970–81	4.3	3.6	4.5	5.4

Source: Ke et al. (1983).

a. Numbers in parentheses are percentages of the annual total.

Table 6.11
Duration Period of Employment in KEPCO and Taipower by Percentage

Number of Years Employed	KEPCO[a]	Taipower[b]
0–5	45.2	29.8
6–10	19.3	19.5
11–15	20.1	18.0
16–20	11.0	9.8
21–25	3.8	11.8
26–30	0.3	5.8
31–35	0.2	2.6
Over 35	0.0	2.7

Sources: Ke et al. (1983); Kim et al. (1983).
a. End of 1981.
b. End of 1982.

old in 1982. There is a possibility that the general level of employee competency has been deteriorating. Much of this phenomenon, however, is attributable to the rapid growth of these utilities. Indeed, even if no one ever left the employ of KEPCO and Taipower, about 25 percent of the employees at KEPCO and Taipower would still have less than five years of experience, and only 38 percent of the employees at KEPCO would have over 20 years of experience.

High turnover of competent employees has been attributed to fewer incentives and benefits and unattractive pay scales compared with those of other enterprises, government or private. Work proficiency, efficiency, and productivity usually improve with length of employment; therefore, a high turnover rate can act as a reverse indicator of these traits.

Besides education, training of employees at all levels has been a useful tool to improve personnel quality. Both KEPCO and Taipower have general as well as specific training programs and well-established training centers. For Taipower, the programs cover off-job and professional on-the-job training, labor and skilled labor training, supervisory training, and special training. The KEPCO program is outlined in Table 6.12. For development of new technology (such as nuclear), qualified engineers and technicians have been sent abroad for training.

Through training, work quality improves, job turnover decreases, and willingness to stay with the company rises. According to KEPCO and Taipower experience, quality improvement has always led to higher efficiency, higher reliability, and better system performance.

Table 6.12
KEPCO Education and Training Programs

Place	Course	Trainees	Scope	Actual 1981 (no. of persons)	Planned 1982 (no. of persons)
KEPCO Training	Basic course	Newcomers	All fields	1,055	1,310
KEPCO Training Center	Developing course	Unskilled employees	Engineers	1,128	1,456
			Administrators	464	715
			Technicians	574	690
KEPCO Training Center	—	Managing staff	All fields	1,934	1,057
Special education facilities	Basic course	Newcomers	Nuclear	221	355
Kori Nuclear Center	Developing course	Unskilled employees	Nuclear	333	437
Other special education facilities	—	Newcomers	Generation transmission computer	142	229
Other domestic education facilities	Graduate school and other	—	—	347	283
Overseas training	Contractual training	—	Power development and other	102	139
	KEPCO planned training	—	Power development and other	41	194

Source: Kim et al. (1983).

Management Structure

The general framework of the management organizations in KEPCO, Taipower, and Tokyo Electric Power Company (TEPCO) are similar, as they are throughout the power industry. (The experience gained from TEPCO is used here to represent the nine Japanese electric power companies.) KEPCO, Taipower, and TEPCO all establish the head office, branch offices, and work site operations as the three business operation organs. The head office is mainly in charge of managerial functions among various business activities. As the generalizing organ of the company, it formulates management goals and policies, designs long- and short-term plans, and plays the role of management and control organ for the branches and work sites.

The branch offices in TEPCO are the intermediary generalizing organs for the entire enterprise. They formulate a plan of execution in concrete terms for their respective control areas in accordance with the management goals and policies and with a basic plan laid down by the head office, and administer and control the production and sales activities of their respective subordinate organs. Unlike TEPCO, in KEPCO and Taipower the branch offices are basically district sales offices in charge of sales, customer service, and maintenance of the secondary distribution systems in their respective regions. Sales offices, electric power offices, power stations, and construction offices in all three enterprises serve as the work site operation organs for carrying out production and sales operations.

Department differentiation. Although generalizations are difficult, the three electric enterprises take the form of ''line'' and ''staff'' organizations. Each department of the head office is a staff department that assists top management in formulating goals, policies, and plans according to respective functions and provides services to the line organs of branch offices, power offices, power plants, and construction offices.

The staff system is instituted in parallel with the line system of business operations. The head of the line (the president) receives staff member assistance and advice on every aspect of business activity, which is based on specialist knowledge and experience. Whereas the line organs in TEPCO report directly to the president, those in KEPCO and Taipower report to their respective head office departments. (In this regard, the management structures of both KEPCO and Taipower may be considered more centralized than that of TEPCO.) Accordingly, the sales and production departments of the head office in KEPCO and Taipower can be regarded as staff departments, although they play a line role with regard to some matters.

Within centralized organizations, top management personnel are often unable to allocate sufficient time for long-term decision making and management policies and strategies because of daily work pressures and heavy work loads. Expansion of management means increasing the number of administrative tiers, hence longer communications channels, which makes it difficult for top management decisions and orders to be made quickly and with validity. Various rules and regulations, promulgated for the purpose of controlling subordinate organizations from the head office, in fact weaken the individual employee's automony in making decisions and taking initiatives. The ensuing discouragement presumably inhibits the exercise of ingenuity and managerial creativity that could enhance the adaptiveness and development of the organization. Centralization further discourages the development of professional managers because, under circumstances where decision-making authority is held solely by top management, low-level managers cannot adapt and prove themselves in order to advance in the organization.

After considering the negative aspects of centralization, KEPCO and Taipower in recent years have made an effort to delegate some of the powers from the head office to field offices through computerization and standardization. In particular, in Taipower's head office general managers always include suggestions and opinions of branch general managers in their reports to the vice president or president. On the other hand, general managers of TEPCO's head office departments, who are members of the staff, are authorized to instruct other department general managers or branch general managers directly. As a result of sanctioning this "functional instruction," a department general manager may virtually exercise line authority, and from this vantage point the department general manager's role will assume a line aspect.

The three electric enterprises have also adopted the functional department system as their form of organization. Whereas divisional differentiation is by region or product, functional differentiation is by function such as sales, production, and services. In the functional department organization, the work is sorted according to its nature; no individual is burdened by several kinds of work at the same time and is able to concentrate on one type only. This characteristic has contributed to stability and specialization of management in KEPCO, Taipower, and TEPCO. On the other hand, vertical communication and horizontal coordination are difficult when problems arise, due to the sophisticated job specialization that separates departments. Problems tend to be solved not from the "macroviewpoint" of the company but from the "microviewpoint" of the specialized department. Because of the division of responsibilities, no one department can be called on to give a companywide perspective. Furthermore, as coordination among departments becomes complicated and the matter at hand is a concern of two or more departments (as is often the case), an upper stratum is forced to solve the problem, which in turn may result in unwanted overconcentration or centralization of management power. Due to the growing complexity of the internal control system, executive functions are greatly weakened. When the number of strata increases and the organization expands in space (regions), decision making at an upper stratum and communication of decisions to the lower strata tend to slow. The flexibility of business activities at the front will then be lost, and decisions may not adequately reflect regional conditions.

Stratification. The business activities in each of the three electric enterprises are divided into the functions of general management, departmental administrative, and work. The management organization corresponding to these functions is stratified into the top management stratum, which establishes basic management policies and plans; the administration stratum, which implements planning and supervises; and the working stratum, which carries out the operational work.

Table 6.13 compares KEPCO, Taipower, and TEPCO employment by organizational strata. The proportion of the top management stratum to total number of employees ranges from a high of 0.08 percent in TEPCO, to 0.06 percent in Taipower, and to a low of 0.04 percent in KEPCO. Specifically, the number of persons in top management in TEPCO is 27, 14 of whom serve not only as members of the board of directors but also as general managers of the head office departments. This rather large number of executives is deemed appropriate and desirable because maneuverable strategic decision making by top management becomes extremely important in times of rapid change, especially when technological innovation advances quickly. The administrating stratum (middle management) as a proportion of total employment is highest in TEPCO (6.7 percent), followed by KEPCO (4.6 percent), and lowest in Taipower (3.6 percent). Conversely, the proportion of the working stratum is highest in Taipower (96 percent), followed by KEPCO (95 percent), and lowest in TEPCO (93 percent). The high proportion of the working stratum in Taipower may be attributed to the fact that most of the jobs for meter reading and billing are carried out by Taipower employees. For instance, in 1983 the number of service company employees contracted for meter reading and billing by KEPCO was 2,990, whereas Taipower contracted only 960.

Flexibility. With growth in the scale of the organization and fractionalization of functional departments, organizations tend to become rigid. To give flexibility to both structure and operation, the three electric enterprises institute various project teams and committees. When an overall goal or task related to several departments or sections of the head office is to be accomplished by focusing energies in a short time, experts from the departments or sections are organized into a project team on a temporary basis to work in an intensive and flexible manner. In KEPCO, Taipower, and TEPCO, there are many formal and informal task forces, and their duration may be long or short.

On the plus side committees have excellent coordinative functions, ease in securing stability, and persuasive power in having their judgments accepted. On the minus side, they lack originality in judgment, they obscure responsi-

Table 6.13
International Comparison of Employment by Organizational Strata

Stratum	Head Office Number	Head Office %	Nonhead Office Number	Nonhead Office %	Total Number	Total %
Managing stratum (top management)						
KEPCO	9	0.6	0	0.0	9	0.04
Taipower	16	0.8	0	0.0	16	0.06
TEPCO	27	1.0	0	0.0	27	0.08
Administrating Stratum						
General manager						
KEPCO	21	1.3	71	0.4	92	0.4
Taipower	22	1.1	93	0.4	115	0.4
TEPCO	52	1.9	56	0.2	108	0.3
Deputy general manager						
KEPCO	32	2.0	90	0.4	122	0.6
Taipower	45	2.1	119	0.5	164	0.6
TEPCO	190	6.8	859	2.5	1,049	2.8
Manager						
KEPCO	167	10.5	604	3.0	771	3.6
Taipower	151	7.1	539	2.1	690	2.6
TEPCO	269	9.6	1,086	3.1	1,355	3.6
Subtotal						
KEPCO	220	13.8	765	3.8	985	4.6
Taipower	218	10.3	751	3.0	969	3.6
TEPCO	511	18.3	2,001	5.8	2,512	6.7
Working stratum (general service personnel)						
KEPCO	1,363	85.6	19,356	96.2	20,719	95.4
Taipower	1,883	88.9	24,010	97.0	25,893	96.3
TEPCO	2,252	80.7	32,684	94.2	34,936	93.2
Total employment						
KEPCO						
Number	1,592	100.0	20,121	100.0	21,713	100.0
Percent of total	(7.3)		(92.7)		(100)	
Taipower						
Number	2,117	100.0	24,761	100.0	26,878	100.0
Percent of total	(7.9)		(92.1)		(100)	
TEPCO						
Number	2,790	100.0	34,685	100.0	37,475	100.0
Percent of total	(7.5)		(92.5)		(100)	

Note: The installed generation capacities of KEPCO, Taipower, and TEPCO are 14,190 MW, 15,231 MW, and 35,630 MW, respectively.

bilities, and they spend much time and money. Accordingly, the three electric power enterprises establish various committees under the following conditions, with due consideration given to their pluses and minuses: when interdepartmental coordination is required to consider the formulation of a general management plan; when an interdepartmental affair closely involves various departments or sections and advance arrangements would ensure smooth execution of the business; when a fair judgment is to be made on such matters as reward or punishment and adoption or rejection of proposals for improvement; when laws and regulations require establishment of a committee; and when the establishment of a committee is judged indispensable to a survey, plan, or research.

If left to their own initiative, committees would grow in number needlessly. Furthermore, if their operation is not appropriate, their effects and responsibilities would be obscured, and the efficiency of the committee as well as the overall efficiency of the organization would be negatively affected. Accordingly, each electric enterprise is prudent in handling the committees by providing standards for their establishment and operation, or by stipulating provisions for the institution of a specific committee. In particular, TEPCO tries to rationalize the operation of committees according to their intent or nature by distinguishing them from other organs involved in planning, consulting, coordinating, or liaison. There are nine main standing committees in TEPCO, six in Taipower, and seven in KEPCO.

In addition to the various project teams and committees, TEPCO has created assistant positions that are usually referred to as being "in charge." Whereas KEPCO utilizes only a limited number of specialists at the head office, TEPCO uses assistant positions extensively. Rapid changes in and out of the country have brought about a remarkable situation where head office department general managers and branch general managers cannot cope with the duties of departmental management without excellent and thorough knowledge and technological ability in special areas such as environment and safety. It is therefore necessary for department general managers and branch managers to have assistants working in these areas. The managers would thus be free to expand their spheres of activity. The assistants may be given some predetermined, specific duties, such as environment and safety, or various ad hoc tasks. In any case, the arrangement gives flexibility to changes in quality or quantity of duties and allows the posting of individuals according to capability and experience. Moreover, since the content of an assistant's specific duties is often common to duties performed in all departments or sections, the arrangement is expected to particularly help the assistant coordinate interdepartmental or sectional activities, thus making the entire organization more dynamic.

In addition to the above arrangements, TEPCO institutes coordinator positions to carry out general stafflike functions in a deliberate and expert manner.

Head office. In both Taipower and TEPCO, the chairman is the chief executive officer presiding over the board of directors, and the president is the chief operating officer, exercising general control according to the decisions of the board. The chairman in TEPCO also presides over the Council of Management Policy, which establishes the company's medium- and long-term management strategies. Unlike Taipower and TEPCO, the responsibilities of the chief executive officer and chief operating officer in KEPCO are retained by the president. As Figure 6.4 illustrates, KEPCO is structured such that 21 departments of the head office are under the responsibility of one of seven division vice presidents, each reporting to the president. The division vice presidents are responsible for the following areas: planning and administration, general affairs and accounting, procurement, power generation, transmission and substations, business and service, and power plant construction. The executive vice president assists the president in the company's overall management as an executive staff member.

In Taipower the eight division vice presidents share the supervision of the twenty-two departments of the head office (see Figure 6.5). The eight divisions are: nuclear operations, nuclear construction, systems operations, power development financing, design and construction, sales and materials supply, general planning and personnel, and general administration. The nuclear operations and nuclear construction divisions are grouped together under the nuclear energy group, which is supervised by the executive vice president. Accordingly, the two vice presidents in charge of the nuclear operations and nuclear construction divisions report to the executive vice president, who then reports to the president, but only in the nuclear field. The other six vice presidents report directly to the president. In Taipower there are five chief engineers performing a staff function for the president and vice presidents.

As Figure 6.6 illustrates, 29 departments of the head office in TEPCO are under the supervision of one of four executive vice presidents and seven managing directors, each of whom reports to the president. In considering the vital role of nuclear power in future power development, both Taipower and TEPCO have grouped all nuclear-related departments under the supervision of a single executive to achieve maximum nuclear management efficiency.

As Table 6.13 shows, head office employment as a proportion of total employment ranges from a high of 7.9 percent in Taipower to 7.5 percent in TEPCO and to a low of 7.3 percent in KEPCO. Table 6.14 compares the number of supporting staff of the presidents of KEPCO, Taipower, and TEPCO. The number of executives supporting the president is highest in TEPCO (25), followed by Taipower (14), and lowest in KEPCO (8). If advisers and coordinators are added the totals are 33, 14, and 11, respectively. The large number of supporting executives in TEPCO corroborates the critical role of maneuverable strategic decision making by top management in adapting promptly in times of rapid change. In addition to the supporting executives, advisers, and coordinators, each department of the head office is a staff department that assists top management in formulating goals, policies, and plans according to its respective functions.

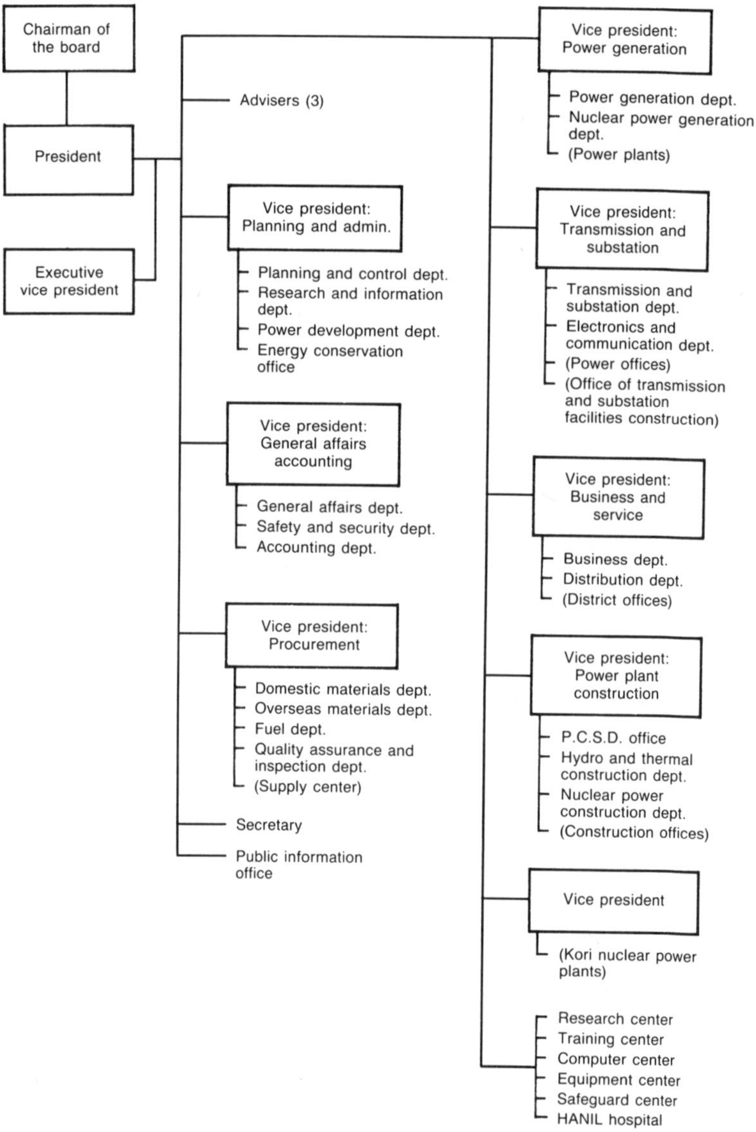

Figure 6.4. KEPCO management structure. Parentheses indicate that facility is not located at main headquarters.

The average number of employees per head office department, as presented in Table 6.15, ranges from 96 (TEPCO and Taipower) to 75 (KEPCO). Whereas the average number of general managers per head office department is 1 for both KEPCO and Taipower, it is 2.3 for TEPCO (including the "in charge"

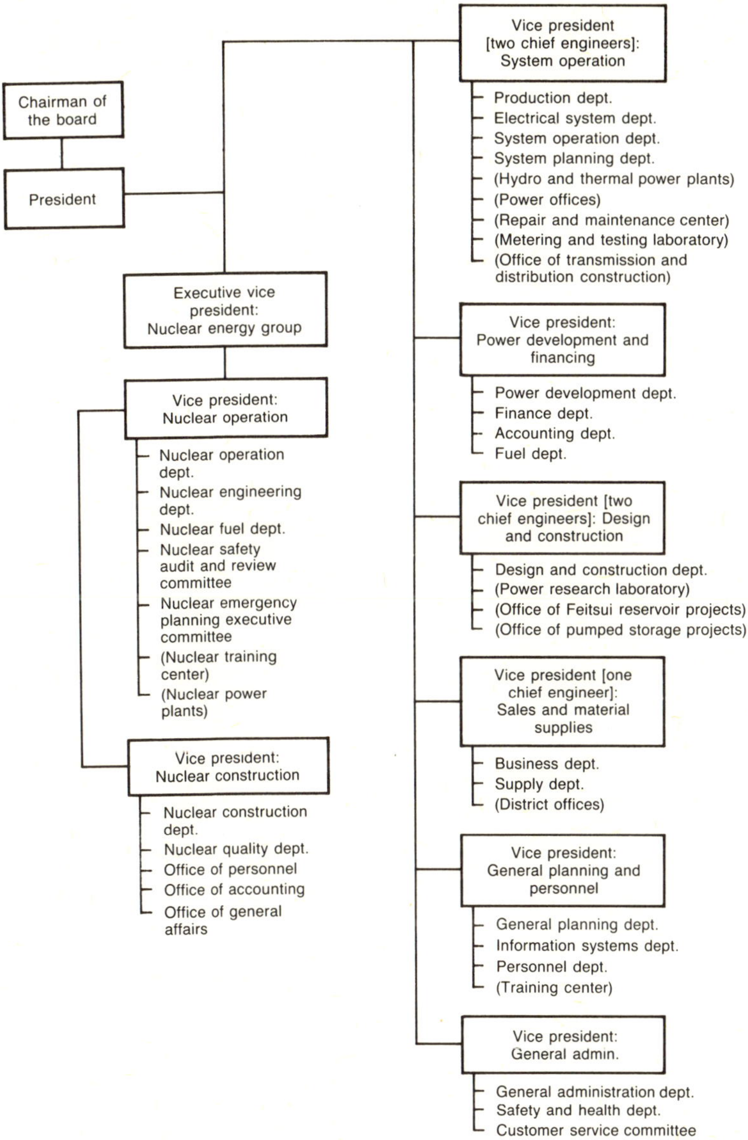

Figure 6.5. Taipower management structure. Parentheses indicate that facility is not located at main headquarters.

assistant positions of the general manager level). On the other hand, the average number of deputy general managers for the head office department is highest in TEPCO (6.6), followed by Taipower (2.1), and lowest in KEPCO (1.5). The deputy general manager is the assistant to the general manager and

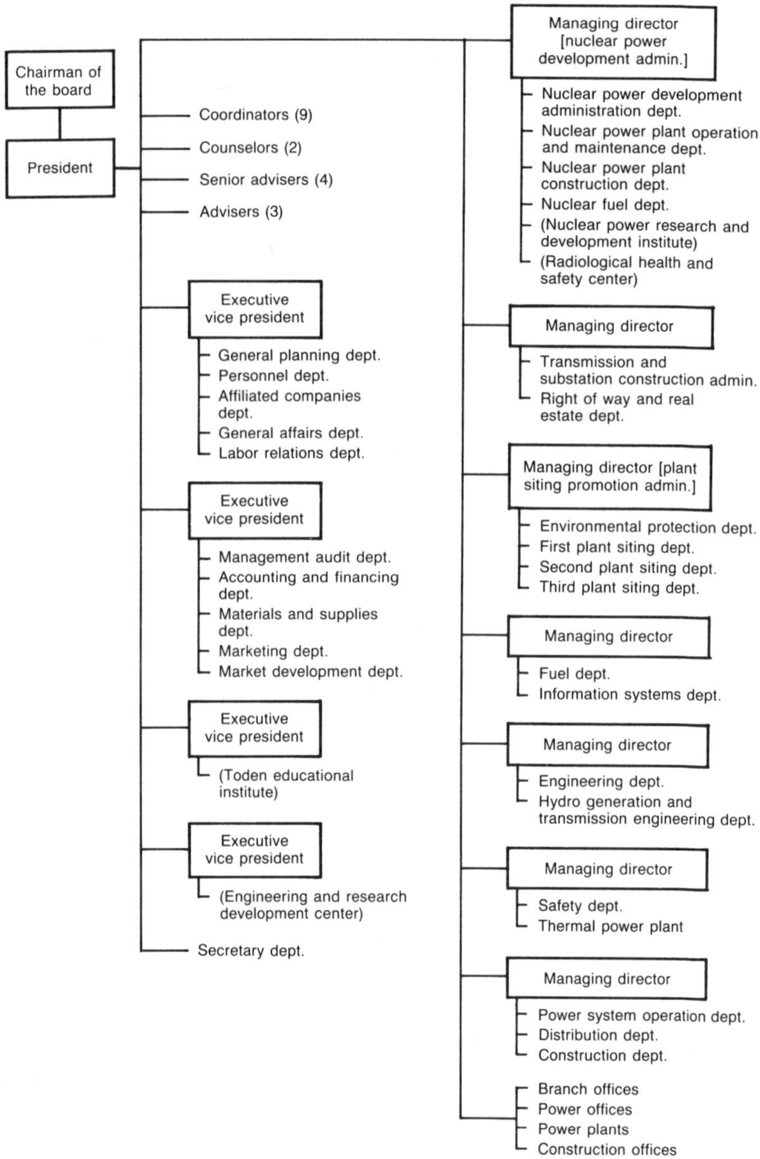

Figure 6.6. TEPCO management structure. Parentheses indicate that facility is not located at main headquarters.

Table 6.14
International Comparison of Supporting Executives for the Presidents

Type of Executive	KEPCO	Taipower	TEPCO
Executive vice presidents	1	1	4
Vice presidents (managing directors)	7	8	7
Directors	0	0	14
Chief engineers	0	5	0
Subtotal	8	14	25
Advisers	3	0	0
Coordinators	0	0	9
Total	11	14	33

a potential candidate for the general managership. Thus the existence of more than two deputy general managers per department represents an important personnel inventory program.

In TEPCO the general planning department of the head office plays the role of "control tower" for the company's overall management. The general planning department serves as a secretariat for the board of managing directors. Also, the general manager of the general planning department holds a director-level position and is a member of the board of directors. He also serves as an executive secretary for the Council of Management Policy whose members are the chairman, the vice president, the four executive vice presidents, the chairman of the auditors, and the counselor.

TEPCO institutes the Council of Management Policy as an organ superior to the board of managing directors. It has three subordinate subcoun-

Table 6.15
International Comparison of Number of Employees per Head Office Department by Functional Position

Type of Employee	KEPCO	Taipower	TEPCO
General managers	1.0	1.0	2.3
Deputy general managers	1.5	2.1	6.6
Managers	8.0	6.9	9.3
Other employees	64.9	85.8	77.7
Total	75.4	95.7	95.8
Number of head office departments	21	22	29

cils: integrated technology development, integrated systems, and integrated nuclear policy, each headed by an executive vice president. In recognizing the necessity of integrating ''top-down'' and ''bottom-up'' oriented management for the enterprise's dynamism and the revitalization of management organization, 10 working groups are placed under the three subcouncils: cost-reduction technology, new energy technology development, technology development policy, information systems, plant automation, teletransmission systems, integrated nuclear policy, nuclear cost-down, nuclear fuel cycle, and the new reactor types working group. Each of the 10 working groups is headed by a general manager of the head office department and consists of 10 deputy general managers and managers. The function of the Council of Management Policy it to perceive the strategic management issues that TEPCO faces in a rapidly changing business environment and make strategic management decisions on the company's basic management policy direction. The 10 working groups are then instructed in this direction through the three subcouncils. The explicit implementation policy options and ideas will be extensively discussed by each of the working groups, thereby mobilizing all existing expertise within the head office departments. The implementation plans are then fed bottom-up through the three subcouncils to the Council of Management Policy. After screening the plans, the council conveys them to the board of managing directors, which is responsible for implementing the policies.

CONCLUSION

In looking back on the past 20 years, the electric power industries in China (Taipei), Japan, and Korea smoothly and successfully carried out interfuel substitutions as national goals, first away from coal and later away from oil, and achieved technological progress in electricity production. This was possible during a period of rapid economic growth, high load growth, and, up to 1973, cheap fuel prices. In particular, between 1960 and 1980 KEPCO accomplished the fastest technological progress in terms of thermal efficiency, rates of transmission and distribution losses, labor productivity and capital efficiency, thereby considerably reducing the technological lead that the Japanese utilities and Taipower had in 1960.

In planning over the next 20 years, the utility management in all three countries, although to varying degrees, is faced with the harsh reality of carrying out interfuel substitution even further away from oil, and at the same time maintaining the efficiency of the power industry. It is also evident that the three countries face excess generating capacities and yet have shortages of peaking services. Since most capacity additions through the late 1980s are already under construction, substantial excess capacity is likely to persist in the three countries through the decade.

Financial difficulties seem likely for all three utility systems in the immediate future. This is probably primarily due to excess baseload capacity resulting from slower load growth and lower load factor, higher interest rates, and a higher proportion of capital-intensive units in newly added generating facilities. In the long run, however, a higher proportion of nuclear generation will substantially reduce fuel costs, thereby improving the financial situation in the three countries. One of the most critical management issues facing these electric utilities is how to reduce generating cost until the planned nuclear capacity is achieved, taking load management into account. Excess capacities can be economically justified during the 1980s, because new baseload generation is cheaper than the fuel costs of existing oil-fired plants. As the utilization of oil plants drops to a capacity factor of 30 to 40 percent, however, further excess capacity will be a substantial burden on the economy, especially in view of the high capital costs of nuclear and coal plants. In the case of Taipower, in particular, the capacity factor of oil-fired plants has been much lower than 10 percent during the 1980s, which means that excessive substitution of coal and nuclear baseload generation for existing oil-fired plants has occurred.

Restructuring the tariff rate system to adapt the load characteristics to those of the generation mix and construction schedule delays may be a very important issue in the near future. Critical to the introduction of more effective time-of-use rates and special tariffs for interruptible power are special sales promotions, other load management techniques such as heat pumps for cooling and heating, hot and chilled water storage techniques, and macroeconomic interactions with the electric power system. In the long run, power-intensive industries such as chemicals, steel, and nonferrous metals may not have comparative advantages in the three countries. In the short run they can act as "sponges" to soak up the excess electrical energy available in the off-peak periods.

Although financial soundness and production costs are of utmost importance to electric utility management, both the quality of personnel and the characteristics of management organization are the most critical factors affecting management efficiency. According to the past experiences of the Japanese utilities, KEPCO, and Taipower, personnel quality improvement has always led to higher efficiency and reliability, and therefore better system performance. Besides education, training of employees at all levels has been a useful tool for improving personnel quality. Through training, work quality improves, job turnover decreases, and willingness to stay with the company rises. Work proficiency, efficiency, and productivity usually improves with length of employment. When compared with other private enterprises, the high turnover rate of competent employees in KEPCO appears to be attributable to fewer incentives, fewer benefits, and unattractive pay scales.

Whereas KEPCO and Taipower are government-owned undertakings, the Japanese electricity supply industry is wholly in the hands of private enterprise. Although government control over utility management is desirable from the perspective of national economic growth, government intervention inhibits the exercise of ingenuity and management creativity by electric enterprises. As mentioned earlier, Korea enacted the Government-Owned Corporation Management Law to provide greater management autonomy to KEPCO. The purpose was to reduce government intervention by establishing a goal-oriented system, and to strengthen management autonomy, thereby creating conditions in which the top management of government-owned corporations could concentrate efforts on achieving management objectives. Along with management autonomy, a performance evaluation system was introduced to reinforce top management's responsibility for the management of the corporation. The improved management system has provided KEPCO with the opportunity to revitalize its management.

REFERENCE

Munasinghe, Mohan, and Warford, Jeremy (1982). *Electricity Pricing: Theory and Case Studies.* Baltimore: Johns Hopkins University Press for the World Bank.

CHAPTER 7

ELECTRIC POWER IN DEVELOPMENT: LESSONS FROM NORTHEAST ASIA

K. R. Smith and Y. H. Kim

Earlier chapters explored the history of the electric power systems in Northeast Asia from 1960 to the early 1980s. Although there have been problems, false starts, and mistakes in estimation, the utilities in these countries have generally been able to meet the prodigious demand for increased quantity and quality of supply that occurred during that critical period of economic growth. In general, however, they face somewhat different challenges in the next decades, partly due to the change in the global energy situation and partly to the different phases of development that these countries have undergone. In particular, environmental and social issues, such as siting controversies, promise to become more prominent.

What of the developing countries that today are situated at the levels of Korea and China (Taipei) 30 years ago? What will their development paths be, and what will be the role of electric power in achieving their goals? These are crucial questions whose answers will affect the material welfare of hundreds of millions of people. Table 1.1 listed some of the middle-income countries that hope to follow Korea and China (Taipei) along a path of rapid economic development in the next 25 years, but in the 1980s their levels of economic development compare with those of the Northeast Asian newly industrializing countries (NICs) in the 1950s.

This chapter extracts from the experience of the NICs of Northeast Asia a few preliminary lessons about electric power systems that may be of relevance to the second- and third-generation NICs. This is done knowing, of course, that the world is a different place today with respect to energy and other factors affecting electric power and development. In addition, the experiences of one country are never completely relevant to those of another, for there are many important differences in the resource base, culture, history, geography, climate, and government that make each country distinct. Nevertheless, such an exercise can be instructive.

Although Japan cannot by any means be considered a NIC since it has had an industrial economy for many decades, it is useful to note the history of the Japanese electric power system. It provides both a realistic point of reference for the path that Korea and China (Taipei) are taking and offers a picture of industrialization that is different from the Western European and North American models where industrialization also occurred somewhat earlier.

Let us start with the lessons that follow directly from the material in earlier chapters and end with a brief discussion of additional important topics that have not been the focus of this particular history. In cases where the countries of most direct interest in this book do not offer appropriate examples, they are taken from other Asian countries.

THE MOST CRITICAL ENERGY RESOURCES

Although generally considered bereft of domestic energy resources, Japan, Korea, and China (Taipei) have shown that they possess large reserves of the most scarce energy resources in the world: sophisticated and effective management, productive and energetic work forces, and the ability to reach and sustain societal consensus. Judicious development of these resources has enabled these countries to provide and maintain power systems during a period of rapid growth and change in their own economies, and of major changes in the world economic and energy systems. Indeed, one could argue that these nongeological assets have served their developers better than have the assets found in the ground in many other countries. This might even include the United States, where a number of utilities have been unable to effectively implement nuclear power plants because of deficits of one or more of these nongeological resources.

Earlier chapters described the critical character of the first of these resources—good management. Not so much has been said of the other two—the work force and societal consensus—because, although important, they reflect conditions outside the direct control of the power sector. Effective management, however, is something that with effort can be brought to bear in one sector even if not yet common elsewhere in the economy.

Management lessons derived from both the successes and failures of this history include:

A vital need in forecasting demand to develop an intimate knowledge of each separate sector as well as uncertainty and not just reliance on overall economic parameters (Chapter 2).

The importance of considering load management early in the development of power systems so that, among other potential problems, large differences do not develop between peak and baseload demand (Chapter 3).

The value of maintaining close management scrutiny and control over indicators of system performance and reliability (Chapter 4).

The importance of management in effective implementation of nuclear power tempered by the recognition that no single technology should dominate supply (Chapters 4, 5).

The value of following indicators of management as well as technical efficiency, and gearing training and organizational structure accordingly (Chapter 6).

In addition to lessons related to management, the earlier chapters provide evidence to support the more general statements made in Chapter 1 about the role of electric power during development (Kim and Smith, 1987).

ELECTRICITY AND ECONOMIC GROWTH

There is much evidence for the importance of electricity in the growth of developing country economies. This evidence is of two kinds: when growth in the electric power sector leads economic growth, and when it trails.[1]

The Leading Economic Sector

As discussed in Chapters 1 and 2, countries that have gone through rapid economic development seem to experience faster growth in energy use than in the economy, and faster growth in electricity than in energy demand. Such was the experience of all three countries of Northeast Asia for most of the period since 1955. The second oil shock in 1979 brought basic changes to the electric supply industries in China (Taipei) and Korea. In addition to lower electricity demand growth, the unbalanced growth of peak and baseload demands was accentuated. With the second oil shock as a turning point, the comparative advantage of China (Taipei) and Korea in international trade shifted against electricity-intensive process industries.

On the other hand, as the standard of living greatly improved, the use of air conditioners and refrigerators increased accordingly. Thus, as noted in Chapters 2 and 3, the summer peak outstripped the winter peak in Japan in 1968, in China (Taipei) since 1972, and in Korea since 1981. Although the trend toward greater increase in electricity consumption with respect to gross national product (GNP) growth was substantially dampened by the oil shocks, the share of electricity in total energy consumption has continued to increase. As a result of the sharp increase in OPEC (Organization of Petroleum Exporting Countries) oil prices, the Japanese economy underwent drastic structural changes in industry, shifting away from electricity-intensive materials.

The shift resulted in the stagnation of baseload growth. Japan has been undergoing a transition from an oil-based to an electricity-based economy, and China (Taipei) and Korea seem to be following suit.

In Northeast Asia, high growth rates in the economy over the last two decades have led to increases of factors of 10 or more in installed electric capacity. As explained earlier, the management and financing of such growth rates are difficult. In addition, utilities in China (Taipei) and Korea were able to greatly improve system efficiency by bringing losses, thermal efficiency, and other technical indicators almost to the level of utilities in Japan and other developed power systems.

Impact of Shortages

In terms of income growth, China (Taipei) and Korea have achieved dramatic results in the last 20 years, and in doing so the electric power sector has grown even faster. Shortages have not generally been a problem, although they were not unknown in China (Taipei), for example.

Other developing countries with different starting conditions, including the two largest, China (Beijing) and India, have not been so successful. Electricity shortages are a daily fact of life and severely curtail production. In a study of Yucheng County (northeast China (Beijing)), for example, it was found that on average 50 percent of customers were waiting for power. In this situation it is necessary to prioritize which groups of consumers will have the highest priority in access to power, or, conversely, who will be cut off first (Baranski, 1985). During the June irrigation period, for example, industry experiences a considerable power cut, and yet it is estimated that only 50 percent of the demand of present irrigation pumps can be met. These pumps, in turn, comprise only about 10 percent of the estimated total demand that would exist if reliable power were available.

In India as a whole it is estimated that there has been considerably more than a 10 percent electricity shortfall in most recent years, much of it seasonal (low water behind dams). This amount may seem manageable until the pattern of shortfall is examined. Because many demands for electricity occur at the same time of day or year, it is often necessary to cut off more than 50 percent of the demand at certain times. For example, during the irrigation season in some areas, demand can increase more than fourfold. In India, the dispatch priority is usually highest to urban residents, second to rural consumers, with industry last. In spite of the dispatch priorities all classes of consumers suffer to some extent, but industry the most. Consequently, in some Indian states industrial electricity shortages approach 50 percent annually. This has a dramatic effect on output. Indeed, it has been estimated that nearly 5 percent of the GNP has been lost from this cause in some years (Pachauri, 1982).

It is important to remember that there are hidden costs to unreliable power other than the obvious short-run marginal loss of production. There are often significant costs to consumers who must provide their own diesel backup or other contingency against loss of power. In addition, there are opportunity costs because of the loss of long-term marginal production from the failure to develop certain industries that may be appropriate for the area in every way except a need for reliable power. India has experienced a "flight of capital" away from some states, particularly in the east, for this reason. Similarly, it has been suggested that the development of the so-called Silicon Valley area of northern California was due partly to the existence of extremely reliable power because of the particular confluence of power grids. Semiconductor manufacture is one of those batch-process industries that is quite vulnerable to power loss.

Owing to a range of interacting factors, including problems with coal production, transport, and quality as well as delays in construction and poor reliability of power plants, these shortages in India are severely damaging. Lack of sufficient revenue to cover costs, difficulties and magnitude of financing, pilferage, and line losses all play roles (Surrey, 1987). As described in earlier chapters, it is just these factors that the Northeast Asian countries were able to bring under control during the 1960s and 1970s.

ELECTRICITY AND INVESTMENT

By far, most international development funding in energy has been devoted to power projects. In the World Bank, for example, power projects typically have accounted for more than 70 percent of all energy lending (Figure 7.1). The World Bank has accounted for about 40 percent of all concessional loans in energy. Other multilateral and bilateral agencies have similarly concentrated on power. The future investment needs of developing countries for power development are even more significant. As Figure 7.2 shows, it is estimated that US$900 billion (1982 dollars) will be needed for energy projects in the low- and middle-income nonoil-exporting countries by the early 1990s; about two-thirds will be needed for power projects, not counting the investments in the basic fuel cycles needed to supply energy to the power sector. This total is about twice the 1982 income of the poorest half of the world's population.

International trade in petroleum usually comes to mind when the relationship between balance of national accounts and energy is mentioned. There are other important trade flows in the energy sector. The principal category includes the capital equipment and construction requirements of power systems. In the early 1980s, for example, the annual oil import bill for developing countries was about US$40 billion, whereas the import bill for capital equipment in the energy sector was approximately US$25 billion (Parikh, 1986). This trade flow is different from that of petroleum since it comprises capital equipment and is most often bought on credit. In addition, the sellers are quite different groups. Power plant equipment comes principally from the larger of the industrial countries, although the NICs—such as Brazil and Korea plus India—have been able to capture a growing fraction of the market.

Largest Capital Output Ratio and Longest Lead Time

The power sector seems to lead overall economic development in the successful cases and to drag down development in the less successful cases, which is a strong incentive for spending a considerable fraction of development cap-

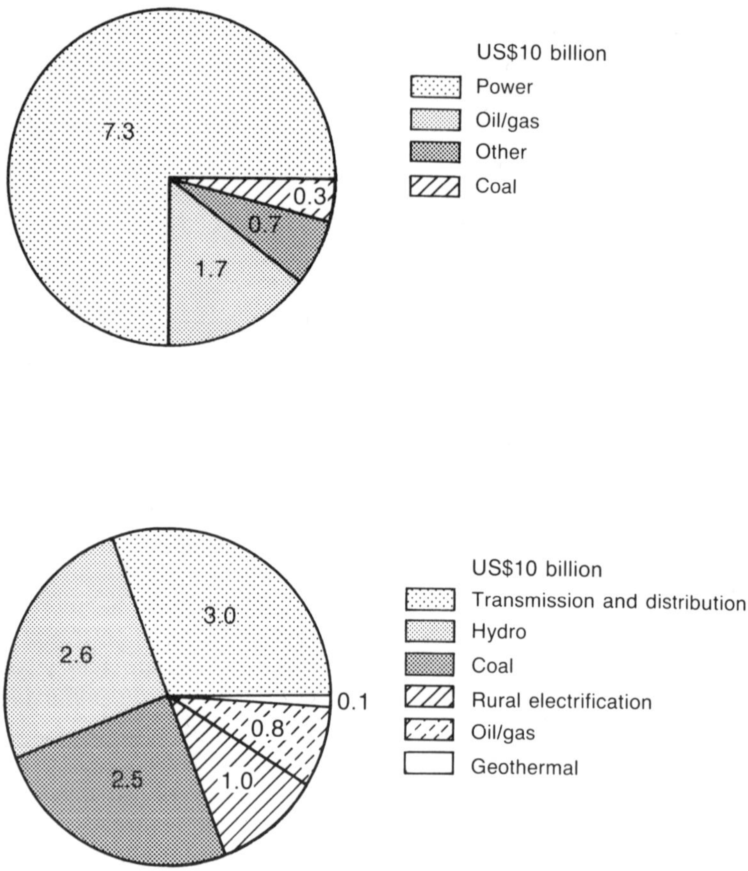

Figure 7.1. Total World Bank lending for (top) energy, 1979–1983, and (bottom) electric power, 1976–1982 (see World Bank, 1983).

ital on power. Indeed, this is the case. Since 1951, for example, nearly 20 percent of the total investment in all development projects in India has gone to the power sector. Typically, more than 35 percent of individual state development plans are devoted to power as well. Even at later stages of development represented by the NICs, the capital requirements of the utility sector are impressive (see Table 1.5).

As discussed in Chapter 6, no other industry has such a large incremental capital output ratio. In other words, the power industry is capital-intensive and labor-unintensive compared to other major industries. This has made it a target for those concerned with the provision of employment-generating industries. On the one hand, many labor-intensive industries such as textiles require electricity. How these two factors balance in any particular combination is difficult to predict without sophisticated econometric studies, which

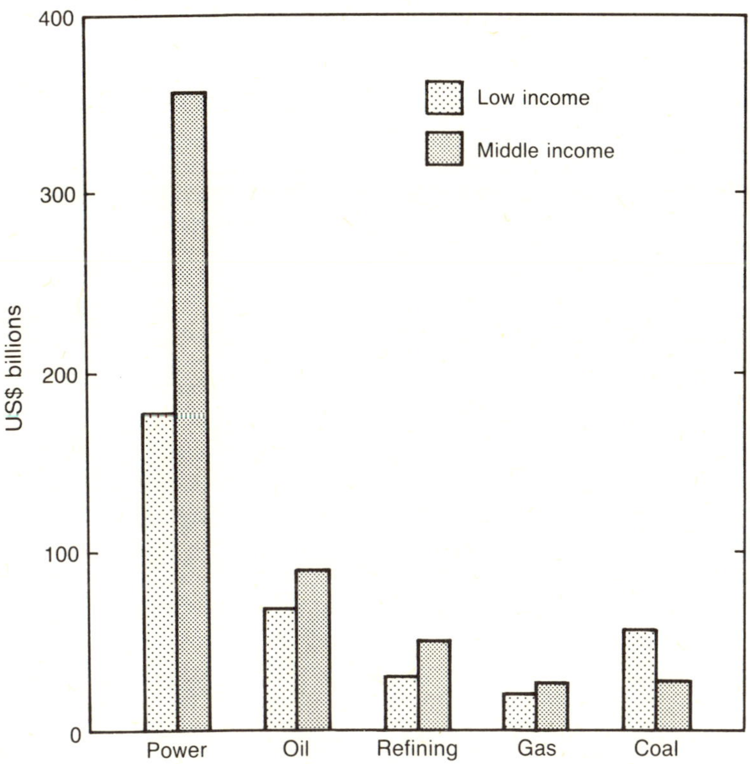

Figure 7.2. Total energy investment for 1982–1992 in low- and middle-income developing countries by sector. Note the dominance by electric power (see World Bank, 1983).

require data that are often unavailable or are constantly being revised. (This issue is also at the core of the social impact of rural electrification and centralized development, discussed below.)

Another characteristic of power systems is their long investment lags. Major nonnuclear baseload power plants take up to six to eight years to bring from conception to completion in most countries, although some countries are able to build plants in less than five years. Nuclear plants can be constructed and readied for use in about five years at the minimum, whereas many projects have taken more than 10 years. Small peaking plants using gas or oil can be built much more quickly, often in less than two years.

Many developing countries obtain much of their power sector investment financing and the capital equipment itself outside the country. This capital need can add significantly to a country's foreign debt. As Table 5.2 indicates, for example, in 1983 in Korea there were about 80 MW of nuclear power being constructed for every billion dollars of GNP, the highest in the world. The total nuclear commitment (in operation and construction) of China (Taipei) exceeded any other country in that year. Adding in the planned power projects at the time would have increased these relative intensities even further.

Since a megawatt of power capacity probably represents close to US$2 million of investment in all parts of the power system, this means that a good fraction of a year's GNP is committed solely to buying power plants and other power system equipment. This is done on credit, of course, but it would seem to commit the country to high economic growth in order to avoid financial difficulties.

The Challenge of Optimum Investment

When the growth of the electric power sector reaches 10 and even 20 percent annually, as it has in many developing countries including the NICs of Northeast Asia, the capital requirements for needed plants can be dramatic. At a 15 percent annual growth rate, for example, the installed capacity increases by a factor of four in 10 years. Assuming that the average time for bringing a megawatt from first conception to full-scale operation is 10 years, a 15 percent growth rate means that at any one time the utility is in various phases of planning, siting, designing, financing, constructing, testing, and training personnel for a system that is four times bigger than any it has dealt with before. It means that the median number of years of experience among the staff will be less than five. Such growth rates also mean that relatively slight changes in demand forecasts can have big effects on the need to plan new capacity. If the true growth rate turned out to be 11.6 rather than 15 percent, for example, the need for capital would be decreased by the amount necessary to purchase a system as large at the beginning (an increase by a factor of three rather than four in 10 years).

This should provide incentives to planners and decision makers in developing countries to examine electricity forecasts carefully. In many cases it may be that a combination of incentives, regulations, and taxes designed to increase efficiency and eliminate large system peaks could avoid the need for expensive new facilities and increase the capital efficiency of the power system. In these circumstances, relatively small improvements in the power sector might free significant amounts of capital for use in other sectors and thus increase the overall growth rate of the economy.

Many countries have been so worried about providing enough capacity during periods of rapid growth that they have paid too little attention to load management. Indeed, as Figure 7.3 shows, the median load factor in Asia-Pacific developing countries has been less than 60 percent. Some observers

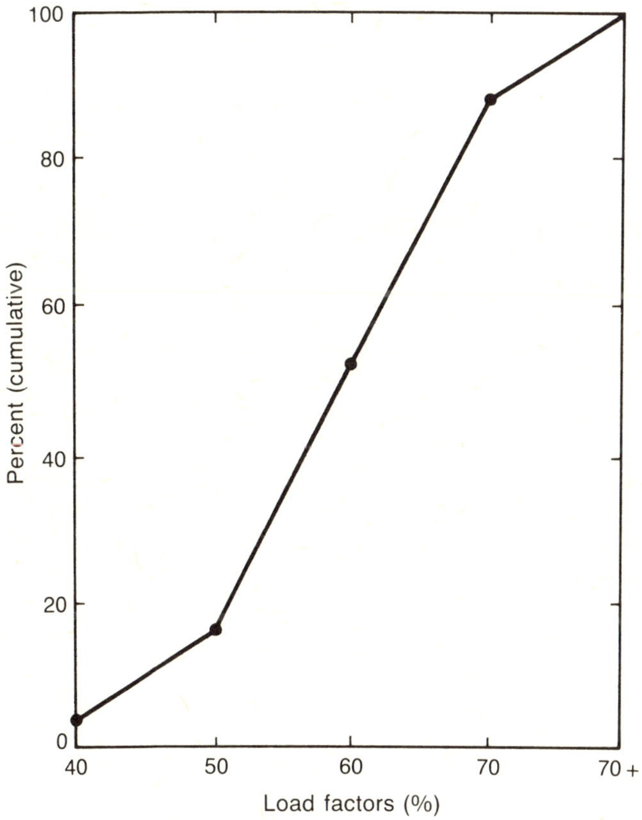

Figure 7.3. Annual load factors for developing countries in the Asia-Pacific region (see ESCAP, 1981).

have noted that these problems have resulted from the influence of too many engineering considerations and not enough economic and financial planning considerations. Consequently, peak loads have sometimes become unnecessarily large. Poor countries tend to have peak loads dominated by irrigation, whereas OECD countries mostly have winter heating and summer air-conditioning peaks. Early in their growth spurts the NICs have peaks dominated by industry, but later residential loads begin to dominate as incomes rise. In China (Taipei) and Korea, for example, as income rose, the summer air-conditioning peak began to dominate just as the growth in electricity-intensive industry began to slow.

As a result of this neglect, increased attention to load management and conservation will likely reap large rewards in many if not most developing countries. At the margin this is cheaper than adding capacity. In addition, as discussed in Chapter 3, advanced technologies for sophisticated load management of both large industrial as well as small residential and commercial consumers are now available. These techniques involve use of radio, power line ripple, or phone communication and cheap reliable microelectronics to greatly increase the rate of interaction between utility and consumer, and therefore, the potential for load leveling.

In addition to low load factors, most developing country power systems have large loss rates, in some cases reaching close to 50 percent (see Figure 7.4). A significant portion of this loss in many areas is due to pilferage and thus might be thought to serve, at least temporarily, some social benefit. Often, however, it is not the very poor who benefit from the pilferage, and over the long term, of course, such losses need to be controlled. The advantage of such control can be large. A system starting with losses of 40 percent, for example, can save itself roughly two years of capacity construction needs by cutting its losses to 20 percent, even if it is growing at the relatively rapid rate of 10 percent. Studies have often shown that there is a higher marginal return on investment in loss control than in new capacity for such systems (Kottewar, 1987). As discussed in Chapter 4, China (Taipei) and Korea were able to steadily reduce losses during their rapid development phases as well as improve the average thermal efficiencies of power plants.

In a manner similar to the potential for loss and load management, the economies of developing countries are often fertile grounds for the introduction of some of the attractive electricity-saving efficiency measures now being developed across the world in response to higher energy prices (see, e.g., Geller, 1987). It is surprising how inefficiently electricity is used in some developing countries. It is hard to dispel the notion, however, that since the total amount of electricity used is relatively small, developing countries need not concern themselves with increasing efficiency. Just the opposite is true, of course, particularly when growth in demand is as rapid as it is in so many developing countries.

Choosing the optimum path between too little and too much investment in the power sector is not easy. Indeed, in the early 1980s much of the world could have been divided into two groups based on which kind of failure occurred. Most industrial countries experienced a surplus of capacity, in some cases in amounts that caused much embarrassment. This resulted from the tendency to project the rapid growth rates in the 1960s and early 1970s into the future and to plan expansion accordingly. Most developing countries experienced the opposite problem, partly because of forecasting problems, but mainly through the inability to quickly arrange system planning, financing, management, and construction. It is instructive to note, however, that the problem of deficit can quickly turn into one of surplus. Both Korea and China

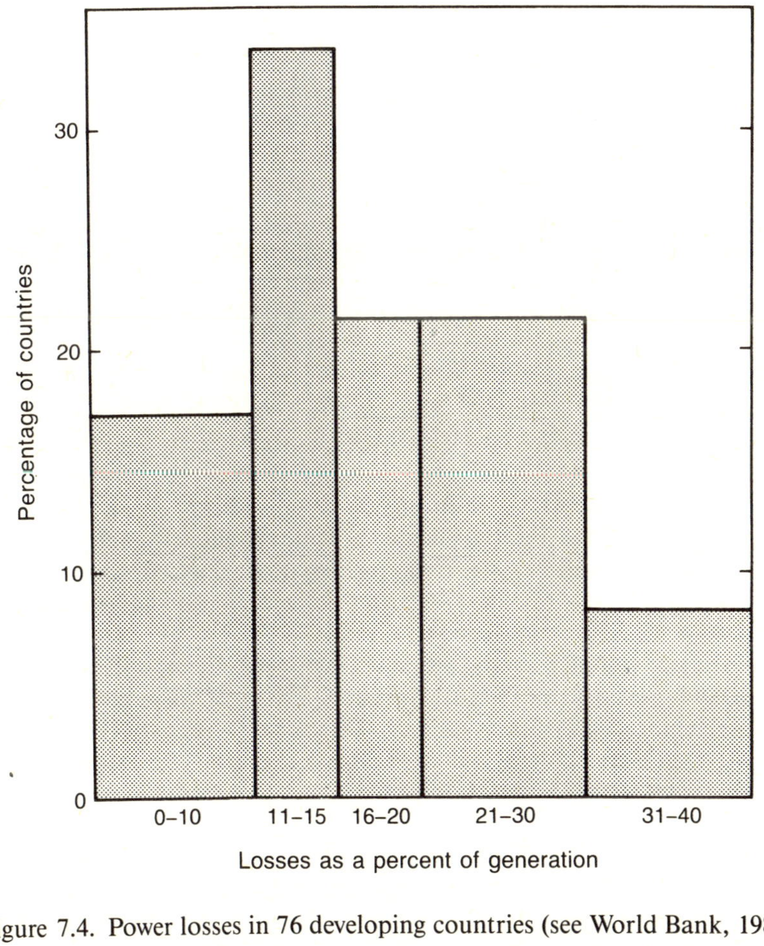

Figure 7.4. Power losses in 76 developing countries (see World Bank, 1982).

(Taipei), for example, experienced significant excess capacities in the mid-1980s and paid a price for not devoting more attention to load management in previous years. They delayed construction of some large baseload plants and began to consider reliance on smaller plants with shorter lead times, as well as concentrating more on load management.

ELECTRICITY: LEADER IN THE POST-OIL TRANSITION

By first approximation, the energy problem has been an oil problem. The central objective of the energy policy of nearly every country in the world is to reduce oil imports or, if an exporter, to reduce domestic consumption so that exports can be sustained. The electric power industry is often the most important agent of change in these efforts.

The Most Homogeneous and Consolidated of Energy Sectors

Changing patterns of direct fuel use among residential, commercial, agricultural, and industrial customers is not easy to accomplish in short time periods. The number and variety of users are large and application of any particular policy tool such as pricing or subsidies often has uneven effects on different consumers, which can create political difficulties. The power sector, on the other hand, is normally consolidated under one or a small number of utilities directly or indirectly controlled by the government. Even when there is more than one, they are not in direct competition and there are enough similarities among their interests to consider them homogeneous for many purposes.

Although the electric power sector is not a final user of energy in a physical sense, it does account for a large fraction of primary energy needs. Typically, 35–40 percent of commercial fuels are used by the power sector in developed countries. Although this might be only 10–15 percent in some developing countries, the growth in demand can be expected to continue to favor electric power, as noted above. Consequently, there is often substantial opportunity to partly shift a nation's pattern of overall demand away from oil by concentrating on the power sector.

This potential illustrates the difference between the controllability of energy used by the consumer versus that used in intermediate production. Typically, economic considerations primarily determine energy use in intermediate production. Relatively little attention is given to any custom, cleanliness, or convenience factors that are not reflected in the price of use. At the final consumption stage, "at the level life is lived" in Robert Heilbroner's words (1959:196), however, such considerations can dominate. Consequently, it is more difficult to effect changes in the final consumption sector through such mechanisms as pricing. Which energy form is used to create the products and services during intermediate production is of little interest to the final consumer. Few people will care, for example, whether a bicycle they buy is manufactured with materials produced with coal or charcoal, whereas they may well care whether their household stove is fueled by coal, wood, or gas. In a similar way, the electricity from coal is the same as that from oil, and thus it is not necessary to expect changes in energy use at the consumer end when shifting fuel sources within the power sector.

The Northeast Asian Example

Energy planners in developing countries are caught between two conflicting demands: providing for rapid growth in energy needs while, at the same time, reducing dependence on imported oil, which historically has been the easiest source to expand. For this reason it is useful to examine cases in which both these goals have been addressed with some success through restructuring the power sector.

Korea and China (Taipei) had incomes and other economic characteristics in the 1950s that are typical of many other developing countries today. Since the 1950s Korea and China (Taipei) have not only experienced significant economic growth, they have also succeeded in reducing the trend toward greater oil dependence. As shown in Figure 1.3 in Chapter 1 and summarized in Figure 7.5, the peak in oil dependence in China (Taipei) occurred a few years after

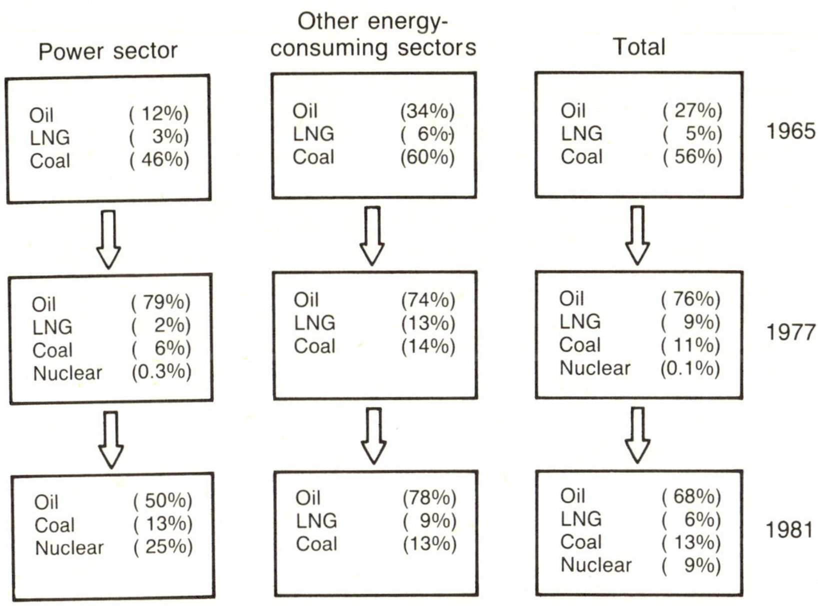

Figure 7.5. Energy transition by end-use sector in China (Taipei).

the first oil crisis, but by 1981 dependence fell more than 10 percentage points overall and by about 25 points in the power sector as new nonoil power plants came on line. In Korea oil dependence reached its highest level in 1979 and has since declined, mostly through the changes in primary fuel requirements of the power sector. In both economies, the ability of other energy-consuming sectors to shift away from oil has been much less impressive. Although part of this success has occurred through the choice to develop nuclear power, which may not be appropriate for many other developing countries, it is significant that coal and liquified natural gas (LNG) have also played key roles along with the important improvements in thermal efficiency and line losses.

On a global basis the potential for additional shifts away from petroleum dependence in the power sector is striking. As Figure 7.6 illustrates, before the oil price drop of the mid-1980s, the World Bank saw a negative growth rate in oil consumption for power during the late 1980s and into the 1990s despite the large growth in the sector as a whole expected during this period.

LEADING ACTOR IN THE EVOLUTION OF THE DOWNSTREAM PETROLEUM MARKET

The ability of the electric power sector to assist in the shift away from oil comes with side effects because of the way this shift interacts with the changing global petroleum market. This causes some uncertainty about the wisdom of moving away from oil too rapidly.

Substitutes for Petroleum

Oil is not used in its natural state but in the form of refined products, whose mix depends on the relative prices and demand of different products, the availability and price of various crude oils, and the cost and flexibility of different refinery configurations. It is only a restatement of the previous section to say that globally most of the successful substitutes for oil have been for power generation, which mainly uses heavy grade products—that is, residual fuel oils. Coal, nuclear, and gas have been the major substitutes, and all, except in countries with gas distribution pipeworks, have principally been used to generate power. Even the most promising of the alternative energy sources—wind, geothermal, biomass, and solar thermal—have their greatest potential in power generation. An exception that has yet to prove its universality is the extensive Brazilian alcohol program. The only new "source" of energy that has shown substantial savings in the light and middle grades of petroleum products has been conservation (increased efficiency), particularly in the transport sector.

As countries all over the world succeed in using substitutes for oil, the result in world markets is a shift in relative demand away from residual oils toward

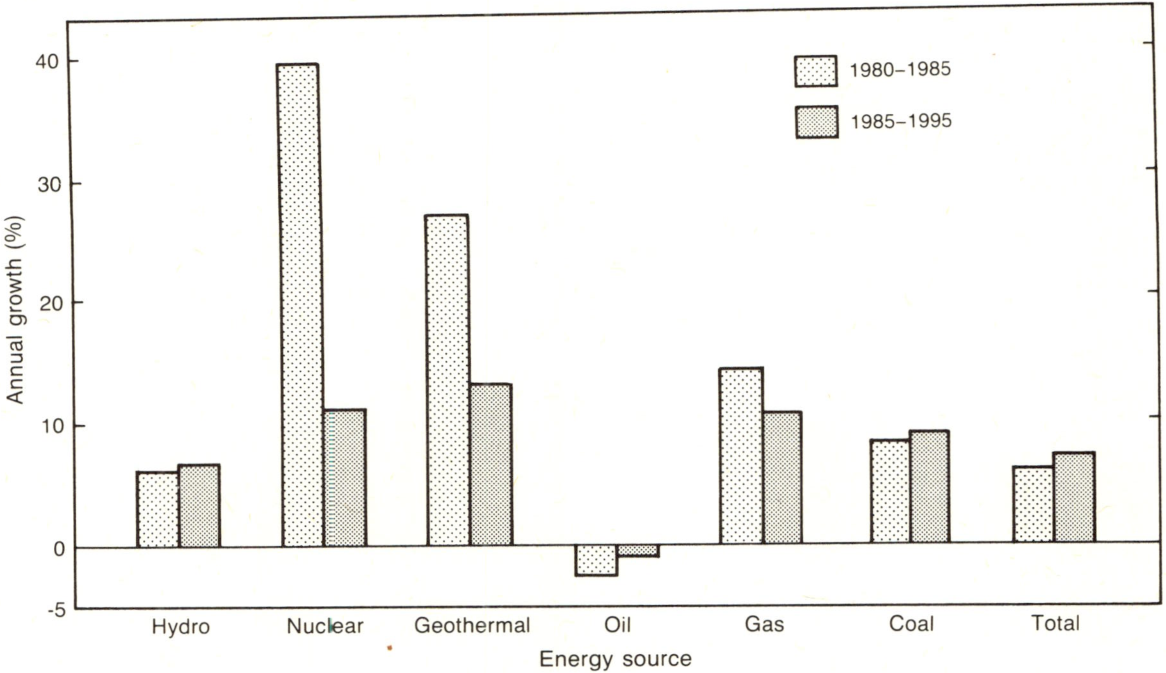

Figure 7.6. Estimated growth in electricity generation by fuel source in developing countries during the 1980s (see World Bank, 1983).

lighter products. Since the greatest growth in energy demand is occurring in developing countries, there is a concomitant increase in demand for the middle grades (middle distillates such as kerosene and diesel). These are the fuels of development, often the first modern (or "commercial") fuels to be used as a society moves away from traditional biomass fuels. In relative terms, the demand for lighter products such as gasoline is remaining relatively stable.

Compared to 1972, for example, the World Bank estimates that the demand for middle distillates will have grown from 30 to 47 percent of total oil demand in the LDCs by 1995 (see Figure 7.7). Heavy grades will drop from 38 to 21 percent, including an absolute annual decline of nearly 1 percent after 1980. Light distillates are estimated to increase from 15 to 18 percent over this period. These shifts are expected to be more dramatic in the relatively fast-growing economies of Asia than they will be in, for example, the poor countries of Africa. The OECD countries will also experience a relative and, possibly, absolute decrease in demand for heavy products along with modest increases in both light and middle distillates (the latter due to increased demand for diesel and jet fuel).

In conjunction with these changes in the pattern of global demand for petroleum products, a significant change is also taking place in global refining patterns. In the 1970s producer countries wrested control over production and crude oil pricing (the upstream market) from the large oil companies. In the 1980s producer countries have been developing their potential for more control over downstream operations as well (refining, transport, and product sales). This can be seen in the expansion of OPEC refining capacity under construction in the Persian Gulf region and elsewhere. This expansion is occurring at a time of great global surplus in refining and while hundreds of refineries are closing in Europe, Japan, North America, the Caribbean, and Singapore—the traditional refining centers. The major Asian oil importers are playing a significant role in this shift (Smith and Fesharaki, 1986).

The Shifting Economics of Residual Fuel Oils

The changes in the downstream market have several important implications for oil-importing developing countries, some relating directly to the power sector. Among these is a need for the means to control and predict the growth in demand not only for petroleum in total but also for each of the major products separately.

The shift away from heavier products presents a country with several options. It can still depend on its own refineries, which in most LDCs are likely to be simple and fairly inflexible facilities that do not have the capability to quickly shift their production to the middle and lighter ends of the product slate. In this case, there may not be any actual decrease in import requirements for crude oil, only an increase in the amount of fuel oil needing export

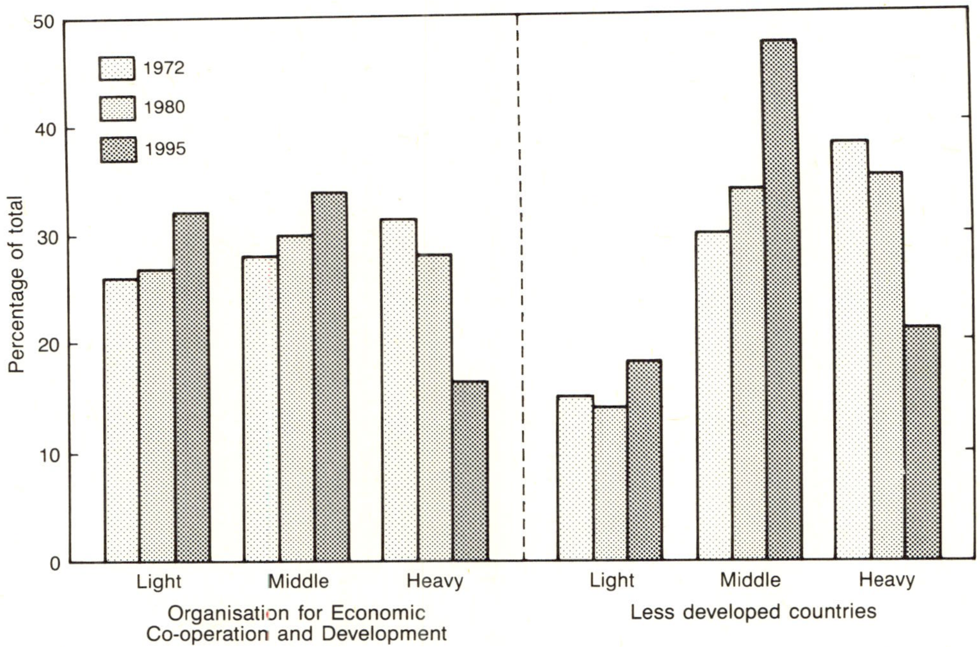

Figure 7.7. Past and forecasted consumption patterns for oil products in developed and developing countries (see World Bank, 1983).

markets. Indeed, depending on the growth in demand for other products, a continued reliance on indigenous refineries can actually lead to growth in crude demand despite an overall decrease in demand for oil. This option is not attractive for these reasons, although it requires no immediate capital expenditures.

Another option is to add to existing refining capacity the capability to crack crude oil into a mix of products more oriented to middle and light distillates. A variant is to buy or lease excess foreign capacity, probably in the entrepôt refining centers of Singapore or the Caribbean. Either way, this option is expensive and carries some risk because of the low refinery capacity factors accompanying global fluctuations in the oil market. In times of oil surplus there is excess capacity, and in times of oil deficit there is lower access to a variety of crude oils. Products, therefore, may be available at the lowest prices from older (already amortized) refineries or those new OPEC refineries with relatively low operating costs, government-subsidized capital, and assured supplies of crude.

A third approach is to increase the imports of petroleum products from the traditional and newly emerging product-exporting refining centers. This is risky because it makes a country even more vulnerable to changes in the demand pattern for products worldwide and to the specific characteristics of refineries built elsewhere by others.

The last option is to implement measures designed to even out the demand pattern across the major product grades in order to match the pattern to existing refineries or a more slowly evolving mix of refineries (a parallel in the oil sector of demand management in the power sector). For example, it may well be in a country's best long-term interests not to move away from oil dependence rapidly if doing so creates a substantial skewing of the product slate.

As a result of the changing downstream market, power capacity planners and dispatchers in national power utilities must consider options based on estimates of the future prices of crude and fuel oils and their ratio. The real price of residual oil dropped and the crude:resid ratio rose as the surplus in residual fuels increased in the early to mid-1980s. Financial analyses in a number of Asian countries have recently shown, for example, that it is often cheaper to retain old oil-fired plants rather than build new coal-fired plants that rely on imports or other long-distance transport of coal. This is a fairly new development in relative prices. If the crude:resid ratio rises much further, in addition, it may well be best from a financial viewpoint to build new oil-fired capacity. This may hold true even when a country has no significant problems caused by the skewing of product demand slates.

This quick overview has shown how important accurate estimates of not only total future oil demand but also the demand within each of the major product groups are to national energy planning. In addition, it is easy to see the value in being able to control the product demand slate so as to reduce skewness. Again, it is the power sector that offers planners the most easily accessible lever to accomplish such manipulations.

THE DILEMMA OF SCALE

Perhaps the most fundamental difference between the economies of China (Taipei) and Korea is their relative scales. The former is characterized by many small firms, whereas Korea has moved toward centralization into a small number of giant firms. This difference is reflected in other differences between the two, including income distribution, the savings rate, and regional balance (Mok, 1986). These differences result from the substantial differences in the historical development of government policies. As discussed in Chapter 6, these characteristics carry over to the utilities, even though both countries have adopted national policies. Other developing countries, however, may not wish to follow the same path as either Korea or China (Taipei).

Indeed, nowhere except in the power sector is the battle over the make-up of optimum development policies so distinct, and within the power sector there is no debate so vigorous as the debate over the proper scale of development. A question confronting planners in many developing countries is (Flavin, 1986): To what extent should power system expansion rely on the extension of central grids connected to relatively large power plants versus decentralized community-size systems?

The answer is not simple and unchanging—for there are many factors and many economies as well as diseconomies to both small- and large-scale facilities. One manifestation of the swings in development thinking has been the relatively recent popularity of the intermediate technology and "small is beautiful" movements, which have emphasized the benefits of scaling technologies according to a broader set of values than just those represented by traditional financial and technical analyses. In particular, they have emphasized the importance of local self-reliance, use of local human and material resources, and the decreased vulnerability to widespread societal disruption that can come with decentralization (Islam et al., 1984).

Another important advantage claimed for small-scale and locally controlled technologies is protection for local cultures from direct internal competition and potential vulnerability to changes in the global socioeconomic system. In addition, it is argued that decentralized systems can better reflect local values and customs. They are also more flexible in that they can more quickly accommodate changes in values.

Such arguments hinge on a distrust of specialization, both in economic and political affairs. Traditionally, however, economic specialization has accompanied economic development. Energy production, for example, was once the task of every family, but in the developed economies it is now the role of specialized workers. This can be seen in the change from wood as the major fuel in Korea to petroleum, a transition that occurred in the 1960s, as Figure 7.8 illustrates. To keep technologies small, close at hand, and labor intensive is to keep the number of specialists to a minimum (Smith, 1987).

An implied distrust of political specialization is also part of such movements as it is among most bottom-up approaches. There is concern that the means of local control over centralized political institutions may be inadequate to the task of representing local values in decision making. There are also advantages to central decision makers of decentralized systems. In making decisions about allocation of resources to a power plant or most other projects, decision makers must make judgments not only about the efficiency (economic viability) of the project but also about its impact on equity (who gains and who loses). Among the advantages of small-scale technologies is that, relative to large-scale systems, it is often easier to apply both efficiency and equity criteria.

The equity criterion is easier to apply because small-scale systems tend to distribute their benefits and deficits to the same group of people. The workers who gain employment, the customers their electricity, and the neighbors

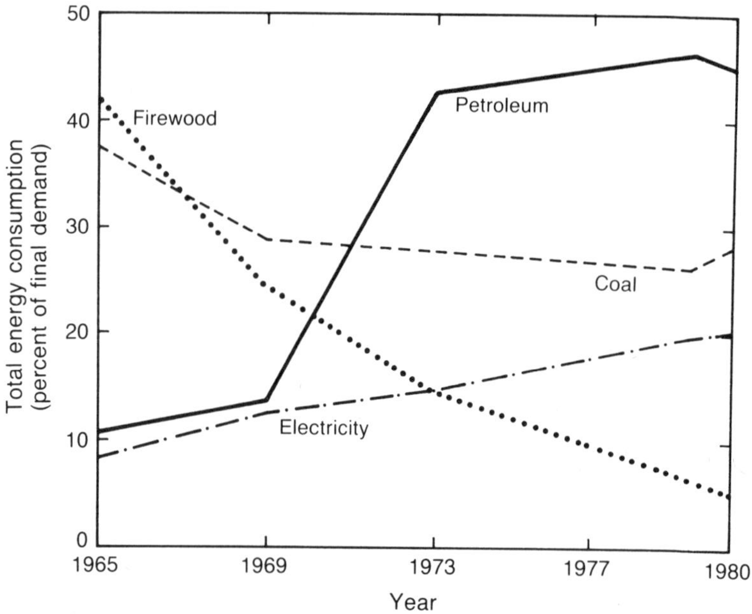

Figure 7.8. Postbiomass energy transition in Korea. Note the replacement of firewood by petroleum products during the period of Korea's rapid rural development in the 1960s and early 1970s (see Smith, 1987).

who experience the environmental hazards are likely to be more congruent, for example, with a small local biomass-powered steam plant than with a large central coal plant.

There are also disadvantages to small-scale systems, such as the apparent economies of scale in construction. History has shown a general, although by no means universal, tendency of decreasing cost per unit for electric power generation, at least up to several hundred megawatts in systems serving fairly densely populated areas. Transmission and distribution costs and losses can have the opposite effect for areas of low population or load density.

It is important to remember that there are also economies of scale in design and management of systems consisting of a few large units rather than thousands of small units fitted to local conditions. Many experts feel that it is these human resources that are in shortest supply in LDCs and, thus, care should be taken to allocate them carefully.

Small systems can also be subject to the "tragedy of the commons" in which the decisions of individuals acting in faith with their individual values can be destructive to the community as a whole. An example might be the damming of a stream in many locations for small-scale hydroelectric power plants where the flow of any one dam would be unable to reach the critical point necessary for power generation. If the damming of the stream were done in concert, however, the critical point could be reached.

There can also be important environmental health advantages to projects done on a large scale. When health insults are produced at a fuel cycle stage that is composed of many small facilities, the negative results per unit of pollutant released may be higher than those from a centralized system. In other words, per ton released, the pollutants from a decentralized power plant can in some cases cause more damage than a ton from a centralized power plant. This is a function of the effectiveness of the pollutants reaching people. Of course, for some types of pollution, such as carbon dioxide, the location of release is not critical. In other types, such as certain types of water pollution, there may be threshold effects that result in less impact per unit of material released from small-scale systems (Smith, 1986b).

Although not a focus of this study, the history presented here shows that the countries of Northeast Asia have generally not attempted to maintain local or even regional foci for power generation. The economies of scale have been seen as so attractive that average unit sizes and other indications of system scale have steadily increased. This has led to some diseconomies as well, which are particularly obvious in the siting of nuclear power plants.

By the mid-1980s, siting was just beginning to become a major public issue in China (Taipei), but not yet a critical factor in Korea. Japan had already faced and addressed this issue by adopting a number of mechanisms to compensate local communities for accepting the siting of large nuclear facilities in their communities. In this manner, some of the equity diseconomies of scale have been remedied. Obviously, however, these issues relate closely to sociocultural and political factors, and as a consequence present difficulties in making generalizations that would apply to other regions.

RURAL ELECTRIFICATION

Rural people in countries that developed in the decades before the major oil price rises turned to petroleum middle distillates (diesel and kerosene) as high-quality substitutes for biomass fuels as well as for energy sources to meet their development energy needs in areas such as tractor power. Figure 7.8 shows this postbiomass energy transition in Korea as fuel wood was largely displaced by petroleum. Because oil was cheap and required relatively little infrastructure, official development efforts concentrated on the provision of the other principal energy form: electricity. This is indeed what happened in Korea, where, as Figure 4.1 (Chapter 4) shows, rural electrification proceeded rapidly in the late 1960s and was essentially complete by the mid-1970s.

Rural electrification expenditures are a large part of the development budgets of developing nations as well as international banks and donor agencies (Figure 7.1, bottom). Since the mid-1970s, however, there has been increasing criticism and cynicism related to many of these efforts. In part this is due to the changing world energy situation. No longer is it safe to assume that rural energy development will take care of itself through the spread of cheap petroleum middle distillates, thus freeing outside development efforts to concentrate on electrification. In a world in which no energy form is cheap, high quality, and abundant, there is a need to look carefully at the relative allocation of scarce development resources in providing fuels versus providing electricity.

Another reason for re-evaluation of rural electrification lies in the expectation by some that electrification would be a way to stem the tide of migration from rural to urban areas. If some of the benefits of life in the city such as lights, radio, and television could be provided in rural areas, the argument goes, the lure of the city would be less. In reality there is conflicting evidence about such an effect. Indeed, the countries with the most complete rural electrification schemes are those with the highest urban migration rates. It thus could be argued that exposure to outside influences and the agricultural unemployment resulting from increased productivity that can accompany electrification could actually be factors favoring urban migration.

These factors have definitely shifted research and policy discussions toward examining energy for rural development in a broader context. In most countries, however, it has not yet led to a significant diminishing of efforts dedicated to rural electrification. This must partly be attributed to the disappointing prospect provided so far by the wide range of indigenous alternative options for energy production that have been investigated. This situation, however, shows signs of improvement. The potential for greater improvement lies in matching planning, management, and labor resources at the local level with ideas and technologies from elsewhere.

In addition, the history of rural electrification is not without significant successes, at least by most measures. The rural electricity cooperative movement in the Philippines, for example, has apparently been able to meet the needs of a wide range of rural people, including many of the poorest. According to studies done in the early 1980s, economic productivity and a series of social welfare indicators improved in electrified areas. The cooperative approach also provides evidence to support the viability of a bottom-up-oriented approach. The Northeast Asian experience also provides evidence of the success of rural electrification, although for the most part this success occurred before the first oil shock through a centralized approach.

Such successes are encouraging and clearly show the benefits that can accompany electrification in an appropriate setting. The fact remains, however, that electricity is nearly useless as a means to directly alleviate the growing shortages of energy for survival. Essentially nowhere in the poorest parts of the rural world will electrical cooking and heating be viable in the next half century.[2] Consequently, rural electrification should not be confused with or seen as a substitute for the immediate need of programs designed to augment supplies for meeting minimum energy needs.

The role of electricity in the early stages of rural economic development might be likened to the role of vitamins in nutrition—vital but not needed in large amounts. The real basic needs for survival, the carbohydrates, are the biomass or other fuels needed to generate heat mostly for cooking and space heating. These are needed continuously and in large amounts, or severe repercussions result. The high-quality fuels of development, analogous to the bodybuilding proteins, have in the past been the petroleum middle distillates. A principal, and as yet unanswered, dilemma of rural development is choosing which fuels will provide the "protein" for development in the future (Smith, 1986a).

This implies that the immediate goal of rural energy development in the poorest countries may be to supply these electric "vitamins," but not necessarily to attempt to supply large amounts. These vitaminlike needs are lighting, communications, and perhaps community or neighborhood refrigeration for medical supplies. This should be done even if the costs per kWh are high. Overall, however, most of the total energy supplied might best be in the form of solid fuels for heat generation and liquid fuels for motive power (the carbohydrates and protein). As development proceeds, a relatively balanced demand from irrigation and small industry as well as residences can be encouraged and, thus, economically met through further electrification.

Poor countries may thus desire to first address other energy needs no longer met without government involvement as they have in the past. Alternatively, vastly different geographical conditions may make electrification much more expensive (as in Indonesia or India). On the other hand, the apparent success of these NICs, particularly China (Taipei), in promoting rural development through electrification should not be ignored. Its success, however, owes much to the broader program of rural development (e.g., promotion of education and small industry) within which it was imbedded (Flavin, 1987).

CONCLUSION

There are indeed lessons for other countries in the past experiences of the Northeast Asian NICs and the example of Japan. Into the 1990s, as Korea and China (Taipei) confront the additional challenges posed by growing domestic environmental concerns about coal and nuclear power, there will be a new set of lessons. The remaining question is the extent to which they can be applied successfully in other countries with differing geographical and social conditions. The information in these seven chapters should be helpful in this process.

NOTES

1. See the speech by Prime Minister Rajiv Gandhi to power planners for a persuasive summary of the importance of electricity in development (Gandhi, 1987).
2. Exceptions to this generalization are, for example, plans in parts of China (Beijing) to use off-peak power for space heating in rural areas.

REFERENCES

Baranski, D. (1985). "Electricity Load Management in Yucheng County." Agricultural Engineering Department, School of Tropical Agriculture, University of Hawaii, Honolulu.

ESCAP (1981). *Optimization of the Utilization of Electricity Generating Plants.* Bangkok: Economic and Social Commission for Asia and the Pacific.

Flavin, C. (1986). "Electricity for a Developing World: New Directions," Worldwatch Paper #70. Washington, D.C.: Worldwatch Institute.

——— (1987). "Electricity in the developing world." *Environment* 29(3):12–15, 39–45.

Gandhi, R. (1987). "Prime Minister's Address." *Urja* 21(1):11–13.

Geller, H. S. (1984). "The Potential for Electricity Conservation in Brazil." Companhia Energetica de Sao Paulo, Sao Paulo.

Heilbroner, R. (1959). *The Future as History.* New York: Harper and Row.

Islam, M. N., Morse, R., and Soesastro, S., eds. (1984). *Rural Energy to Meet Development Needs: Asian Village Approaches.* Boulder: Westview Press.

Kim, Y. H., and Smith, K. R. (1987). "Electric power in economic development: The Northeast Asian experience." In R. K. Pachauri, ed., *Global Energy Interactions,* pp. 621–642. New Delhi: Allied Publishers.

Kottewar, A. B. (1987). "Power losses in distribution network, legal and economic concept." *Urja* 21(1):23–25.

Mok, S. S. (1986). "The macro and micro roads to economic success." *Far Eastern Economic Review* 26 (June):74–76.

Pachauri, R. K. (1982). "Electric power and economic development: The case of India." *Energy Policy* 10(3):189–202.

Parikh, J. K. (1986). "Capital goods for energy development: Power equipment for developing countries." *Annual Review of Energy* 11:417–450.

Smith, K. R. (1986a). "Electric power in developing countries: A brief exposition of its role in the development dialectic." In C. Chagas and U. Colombo, eds., *Energy for Survival and Development,* pp. 243–271. Vatican City: Pontifical Academy of Sciences.

______ (1986b). "Village cooks: The bright and dark sides of small is beautiful." *Environmental Management* 10(1):61–74.

______ (1987). "The biofuel transition." *Pacific and Asian Journal of Energy* 1(1):13–32.

Smith, K. R., and Fesharaki, F. (1986). *OPEC and Asia: The Changing Structure of the Oil Market.* New York: Pergamon Press.

Surrey, J. (1987). "Electric power plant in India: A strategy of self-reliance." *Energy Policy* 15(6):503–521.

World Bank (1982). *Energy Efficiency: Optimization of Electric Power System Losses.* Washington, D.C.: World Bank.

______ (1983). *The Energy Transition in Developing Countries.* Washington, D.C.: World Bank.

Selected Bibliography

Most of the data used in this book were taken directly from the records of the utilities in the three countries: Korean Electric Power Corporation, Taiwan Power Corporation, and, in Japan, principally Tokyo Electric Power Corporation. The data, which have in some cases been updated, are available in the following reports from the Resource Systems Institute of the East-West Center (Honolulu, Hawaii 96848).

Ke, W. P., Lee, K. C., Li, S. C., Lin, E., Pan, P. S., Yand, T. S., and Kim, Y. H. (1983). "The Electric Future of China (Taipei)," RM-83-6.

Kim, E., Kim, H. T., Lim, H. K., Nahm, C. I., Shin, M. C., and Kim, Y. H. (1983). "The Electric Future of Korea," Rm-83-8.

Kimura, T., Kuwahara, K., Sawada, T., Tani, A., Uchida, M., and Kim, Y. H. (1983). "The Electric Future of Japan," Rm-83-7.

Based mainly on these data, a number of related publications have also been prepared at the East-West Center.

Kadir, A. (1984a). "Some Development Aspects of Electric Power in Japan, South Korea, and Taiwan." *Berita PLN* (Journal of the State Electricity Corporation of Indonesia) 39(6):9–14, 19–20.

______ (1984b). "Some Development Issues in the Electric Power Sector in Japan, Korea, and Taiwan." *Pertambangan dan Energi* (Journal of Mining and Energy) 18:45–60.

Kadir, A., and Kim, Y. H. (1984a). "Issues in the Development of Electric Power in Japan, South Korea, and Taiwan." *Energi dan Listrik* (Energy and Electricity) 1:1–15.

______ (1984b). "Northeast Asian Experiences in Electric Power Development," Working Paper WP-84-6. Honolulu: Resource Systems Institute, East-West Center.

Kadir, A., Kim, Y. H., and Uchida, M. (1985a). "Management Efficiency in Electric Power Utilities: Cases of Japan, South Korea and Taiwan." *Energi dan Listrik* (Energy and Electricity) 1:10–27.

______ (1985b). "Management Efficiency in the Electric Power Supply Industry: Experiences in Northeast Asian Utilities," Working Paper WP-85-3. Honolulu: Resource Systems Institute, East-West Center.

Kim, Y. H. (1982). "Korea." In M. S. Wionczek, Gerald Foley, and Ariane van Buren, eds., *Energy in the Transiton from Rural Subsistence,* pp. 159–178. Boulder, CO: Westview Press.

______ (1983a). "Major Issues and Their Policy Implications in the Development of Electric Power Systems: Cases of China (Taipei), Japan, and Korea," Working Paper WP-83-21. Honolulu: Resource Systems Institute, East-West Center.

______ (1983b). "Rational and Effective Use of Energy in Korea's Industrialization." *Energy, The International Journal* 8(1):107–123.

______ (1984). "Interactions among Economic Activity, Energy Use, and Electricity Use." *Energy, The International Journal* 9(9/10):717–725.

Kim, Y. H., and Smith, K. R. (1987). "Electric Power in Economic Development: The Northeast Asian Experience." In R. K. Pachauri, ed., *Global Energy Interactions,* pp. 620–642. New Delhi: Allied Publishers. Also in Working Paper WP-84-3 (1984). Honolulu: Resource Systems Institute, East-West Center.

Rose, D. J., and Wakabayashi, H., eds. (1984a). *Nuclear-Electric Power in the Asia-Pacific Region.* Published as a special issue of *Energy, The International Journal* 9(9/10).

Smith, K. R. (1982). "Nuclear Power in the Asia-Pacific Region." In F. Fesharaki et al., *Critical Issues in Asia and the Pacific: The Next Twenty Years,* pp. 125–181. Boulder, CO: Westview Press.

______ (1986). "Electric Power in Developing Countries: A Brief Exposition of Its Role in the Development Dialectic." In C. Chagas and U. Colombo, eds., *Energy for Survival and Development,* pp. 243–271. Vatican City: Pontifical Academy of Sciences.

Smith, K. R., Rose, D. J., and Kim, Y. H. (1985). "Nuclear Power and Northeast Asia: Its Special Role." Paper presented at the Second Workshop on Nuclear Electric Power in the Asia-Pacific Region. Korean Power Engineering Corporation, Seoul.

INDEX

About the Editors and Contributors

Dr. Manuel Castillo Bonet is a private consultant for electric utility planning and operations in Madrid, Spain. During the preparation of this book, he was a doctoral candidate in the Nuclear Engineering Department of the Massachusetts Institute of Technology.

Dr. Yoon Hyung Kim is a Research Associate in the Resource Systems Institute, East-West Center, and Professor of Economics, Hankuk University of Foreign Studies, Seoul.

Mr. Mitsugu Nakamura is a professor in the Faculty of Economics, University of Tokyo.

Mr. Pu-Sen Pan is former Chief Engineer for Power Development Planning and is now an adviser to the Taiwan Power Company.

Dr. David J. Rose was Professor of Nuclear Engineering at the Massachusetts Institute of Technology and Research Fellow in the Resource Systems Institute, East-West Center (deceased October 1985).

Dr. Kirk R. Smith was Energy Program Leader, Resource Systems Institute, East-West Center (1978–1985) and is now Research Associate in the Environment and Policy Institute.

Dr. Richard Tabors is a lecturer, principal research associate, and Assistant Director of the Laboratory for Electromagnetic and Electronic Systems at the Massachusetts Institute of Technology.

Dr. Oliver S. Yu is Manager for Planning Analysis, Planning and Evaluation Division, Electric Power Research Institute, Palo Alto, California.